Legal information

© 2023
Author and Editor: M.Eng. Johannes Wild
A94689H39927F
Email: 3dtech@gmx.de

The complete imprint of the book can be found on the last pages!

This work is protected by copyright

The work, including its parts, is protected by copyright. Any use outside the narrow limits of copyright law without the consent of the author is prohibited. This applies in particular to electronic or other reproduction, translation, distribution and making publicly available. No part of the work may be reproduced, processed or distributed without written permission of the author! All rights reserved.

All information contained in this book has been compiled to the best of our knowledge and has been carefully checked. However, the publisher and the author do not guarantee the timeliness, accuracy, completeness and quality of the information provided. This book is for educational purposes only and does <u>not</u> constitute a recommendation for action. The use of this book and the implementation of the information contained therein is expressly at your own risk. In particular, no warranty or liability is given for damages of a material or immaterial nature on the part of the author and publisher for the use or non-use of information in this book. This book does not claim to be complete or error-free. Legal claims and claims for damages are excluded. The operators of the respective Internet sites referred to in this book are exclusively responsible for the content of their site. The publisher and the author have no influence on the design and contents of third party internet web sites. The publisher and author therefore distance themselves from all external content. At the time of use, no illegal content was present on the websites. The trademarks and common names cited in this book remain the sole property of the respective author or rights holder.

<u>Caution:</u> **Electricity, especially alternating current and high amperage, are life-threatening. Practical work may only be carried out by individuals who have been professionally trained for this purpose. No liability is accepted if the contents of this book are imitated.**

Table of contents

1 Introduction

Photovoltaics can be understood as the process of converting sunlight into usable electricity. This type of electricity generation – along with other renewable energies, e.g., wind power and hydropower – has experienced strong growth in recent years due to its environmentally friendly technology. Power generation from sunlight using photovoltaic cells (a.k.a. solar cells) can be both off-grid and grid-tied. What these two terms mean, we will learn later. Photovoltaic cells are often colloquially referred to as solar cells, and a photovoltaic system is often referred to as a solar system. However, this umbrella term is also used for solar cells that are used to heat water (solar thermal). However, there is a clear and significant difference between these two systems (photovoltaic and solar thermal), both in terms of construction and operation. In fact, they are completely different systems (unless you use a photovoltaic system in combination with a heating rod to heat water). The only thing they have in common is that both systems use radiant energy from the sun. Photovoltaic uses this energy to generate electricity and solar thermal to heat water.

PV (photovoltaic) systems are widespread all over the world, both on a very small scale, e.g., as a balcony power plant with only one PV module to be connected to one's own household socket, and on a commercial scale, e.g. as a solar farm covering several hectares. With photovoltaic systems, the efficiency of power generation depends on the weather (solar radiation) on the one hand, and on the design of the system on the other. In order to achieve maximum energy yield, a performance analysis should always be carried out.

The performance ratio is one of the most important quality factors here for assessing the performance of a photovoltaic system. This so-called performance ratio is basically the ratio between the possible yield (target) of the installed power of a PV system and the actual yield (actual). But more about that later.

The modules of a PV system are usually fixed (at a certain angle). However, there are now new technologies that allow the PV system to track the sun's path so that it is always efficiently aligned. Fixed systems are usually installed at an angle that allows maximum power generation. The tilt angle of the solar modules depends on the location of the photovoltaic system. For example, if the photovoltaic system is located in the Southern Hemisphere (i.e., south of the equator: e.g., South Africa, Australia, Argentina), a north orientation may be appropriate. This is because in the Southern Hemisphere - south of the Tropic of Capricorn – the sun is in the north during the day (at noon), instead of in the south, as in Europe or the USA. Europe, Canada, the USA, and Mexico are located in the Northern Hemisphere, and a south-facing PV system is suitable here (in normal cases).

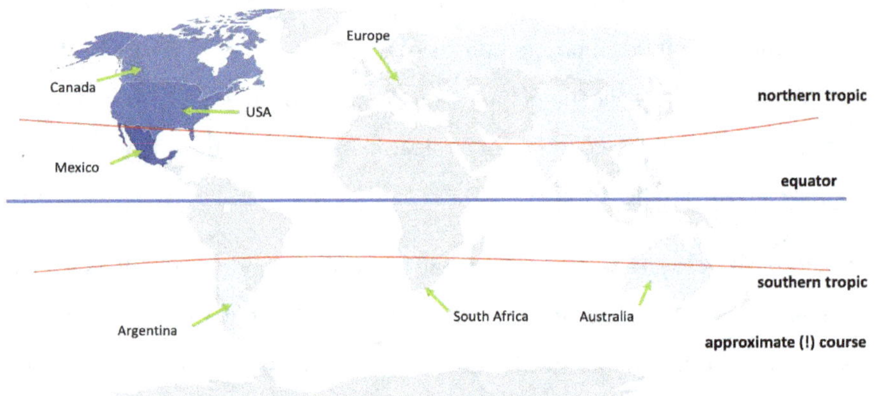

For PV systems with tracking (flexible orientation), a special algorithm is used to move the PV modules along the course of the sun. In this case, the PV modules are normally aligned to the east in the morning (sunrise) and move from the east to the west (sun's course). In this way, the PV modules are optimally oriented towards the sun throughout the day. This contributes to higher electricity generation. In addition, new technologies are also being used in the modules themselves to improve the efficiency of a system and thus increase electricity generation. Here one can mention bifacial solar modules, for example. The special feature of these modules is that they can be irradiated with sunlight not only on one side but can generate electricity on both sides of the module. In a certain arrangement, for example, the sun's rays that shine past the cell onto the ground can also be reflected onto the back of the PV panels and thus also be used to generate electricity.

Since the beginning of the 19th century, most of the world's energy has been generated using fossil fuels and nuclear power. Due to the dwindling resources of fossil fuels and their harmful effects on the environment, it is of enormous importance to drive an energy turnaround with the help of alternative sources for energy generation. As the demand for energy continues to increase while the energy market remains tight, solar energy is seen as one of the most important ways to generate sustainable electricity. For every nation and even for every household, a clean and cheap source of energy is essential for production (consumer goods) and supply (electric stove, refrigerator, ...). This book should help that we as individuals do not have to wait for the political energy turnaround but can actively shape it by meeting our energy needs with the help of the sun. In the following, we will first start with the most important electrotechnical basics for the topic of photovoltaics. Later on, we will take a look at the different types of PV modules, the installation of PV modules, the construction of an on-grid or off-grid PV system, with or without battery storage, all the necessary components, as well as the concrete composition of components for planning your own solar system. Two practical examples round off the content of the book in the final chapters.

2 Fundamentals of electrical engineering and photovoltaics

2.1 Basic terms of electrical engineering

In this chapter, we will discuss some basic electrical engineering terms that are essential for understanding how photovoltaic systems work and how they are calculated. Please note that this can only be a brief introduction to electrical engineering. If you do not have any knowledge in this field, you should consult an electrical engineering reference book beforehand.

Electrical engineering is significantly based on two fundamental physical quantities that are already dealt with in school – namely charge and energy (work). Andre Ampere was the first to discover these properties of electricity, which are used in the form of current and voltage for the analysis of electrical and electronic circuits.

2.1.1 The electric charge

Electric **charge**, measured in Coulombs (C) and described by the letter **Q** (or q), is a physical quantity that has the property of experiencing a force when placed in an electromagnetic field. What does this mean, and what is an electromagnetic field? An electromagnetic field is composed of an electric field and a magnetic field, which are coupled together. It is a kind of state of space or an area where accelerated charges are located. Humans cannot differentially perceive electromagnetic fields with their sense organs, except for the visible range, which everyone perceives as light. Today, it is hard to imagine life without electromagnetic fields. For example, every microwave works with the microwaves of the same name, and every cell phone also works with microwave radiation. The inverter of a photovoltaic system also generates an electromagnetic field.

There are two types of charges: positive (+) and negative (-). Equal charges repel each other, unequal charges attract each other. We come into contact with charges in our everyday lives more often than we would think. Who doesn't know the crackling and disheveled hair when you take off grandma's wool sweater? Or the small electric shock when touching a door handle or a metal part, if the combination between shoe sole and floor covering (e.g., rubber sole and carpet) is unfavorable. The origin for these everyday experiences are charges. Every object has positive and negative charges that are normally in balance. However, friction during dressing or walking shifts this balance of charges, creating electrical voltage. When the hairs become charged as the wool sweater is put on, they either get stuck anywhere (attraction) or appear to float as they repel each other. This happens because of the equal or opposite charge (two equal charges repel each other, two different charges attract each other).

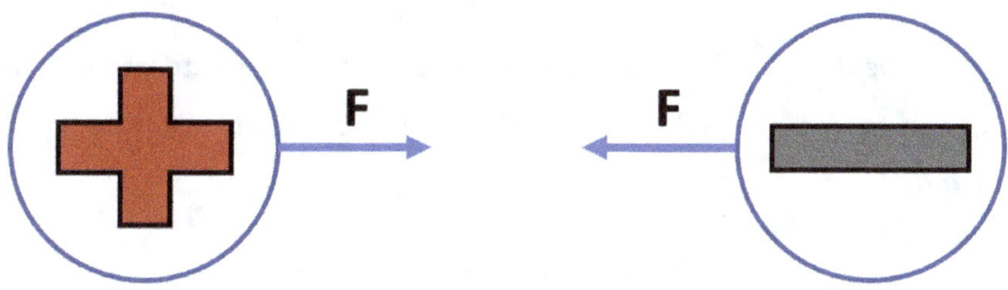

2.1.2 The electrical voltage

Voltage is another fundamental quantity in electrical engineering. The **voltage U** (unit: Volt) can be understood as the energy change (or work W) of a charge in motion. Thus, if a charge of 1 coulomb undergoes a change in energy of 1 joule, this means that there is a change in energy of 1 volt. We also call this **potential difference** (between two points in the electric field). Voltage is mathematically described as:

$$U(t) = \frac{dW}{dQ} = \frac{J}{C} = Volt$$

2.1.3 The current

Electric **current** is defined as charges in motion, or in other words, as: Flow of electrons per unit time. Electric current is denoted by the letter **I** and its unit by the letter A for ampere. Current is mathematically described as:

$$I(t) = \frac{dQ(t)}{dt} = \frac{C}{s} = Ampere$$

Current flow generally depends on the type of conducting material. In conductors, electron flow is due to the flow of electrons, while electron flow in semiconductors is due to the interaction of holes and electrons. But we'll get to that in a moment!

Current can be thought of simply as the flow of charges in an electrical conductor. Voltage, on the other hand, is simply the energy of a charge flowing in that conductor.

2.1.4 The electrical power

In addition to the two basic parameters of voltage and current, we also have to deal with the concept of **power**. For example, you have probably noticed that household appliances such as light bulbs, computer power supplies, microwaves or air conditioners, for example, have the indication 100 W or 250 W. The letter W here stands for watt, the unit of power. **Power P** is defined as work per unit of time. Power indicates how much energy a component supplies or consumes. Power can be described mathematically as:

$$P = \frac{dW}{dt} = \frac{dQ}{dt} \cdot \frac{dW}{dQ} = U \cdot I$$

In everyday life and in the field of photovoltaics, you will often encounter the unit **kWh.** A **kWh** is simply the abbreviation for 1000 watts multiplied by one hour. So, 1 kWh is the energy that a device with a power of 1000 watts absorbs or emits in one hour. To put it even more simply: If a light bulb, with 20 W, runs continuously for 50 hours, it consumes an energy of 1 kWh (20 x 50 = 1000). For this bulb, 20 W means the energy consumption of 20 J in 1 second, and 1 kWh just means the consumption of 20 W of power for 50 hours.

2.1.5 The direct current (DC):

With **direct current** (**DC**), the direction of flow of the electrons (or the polarity) remains the same over time. Nowadays, direct current is mostly limited to low-voltage applications. Why direct current is inefficient for the purpose of transmitting power (e.g., utility pole) is also an interesting topic, but we will not look at it here. Batteries and PV modules also supply direct current. Alternating current, on the other hand, comes from a standard household power outlet. We will look at how the transfer from direct current to alternating current takes place when we take a closer look at PV technology.

2.1.6 The alternating current (AC):

Alternating current (**AC**) varies sinusoidally with time. The following figure shows a sine wave. These types of waves have a **period** and a so-called **phase**. The period is simply the time after which the pattern of a wave repeats. A simple sine function $f(x)=\sin(x)$ has a period of 2π. The term phase describes the displacement of a wave. For example, the red wave in the following figure is shifted $\pi/2$ to the right on the x-axis, so its phase is $\pi/2$. In the formula of the function, this is expressed by the term $-\pi/2$. The **frequency** of a sine wave is the reciprocal of its period. The frequency is measured in Hertz, named after the German physicist Heinrich Hertz. 1 Hz corresponds to one oscillation per second, i.e., 1/s.

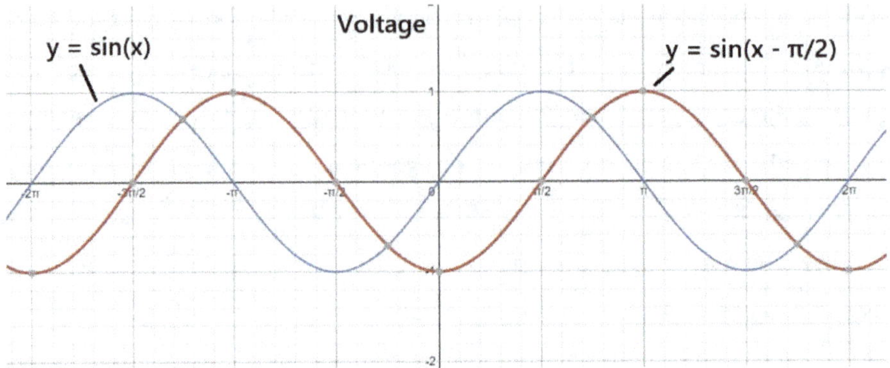

The frequency of the household current is 50 Hz or 60 Hz and varies depending on the region. 50 Hz means that a cycle can repeat 50 times in one second. 50 Hz also means that the AC wave crosses the zero-sequence voltage (x-axis in the coordinate system) 50 times in one second, since the direction of the AC current / voltage changes – as already known. If you decide to use alternating current compared to direct current, you can easily change voltage and current with the help of transformers with only small losses.

2.1.7 Voltage measurement and current measurement using a multimeter

In electrical engineering practice, we often use multimeters as measuring instruments. Multimeters with two terminals can measure voltage, current, resistance, capacitance, and inductance. They can also measure the polarity of transistors and perform a continuity test with them. The continuity test tells us whether a circuit is shorted or not. Multimeters can only measure one variable at a time (like current or voltage). To measure multiple parameters, we need to use several individual devices. The figure below shows a simple multimeter with the different measurement ranges. Depending on what you want to measure, you turn the dial to the appropriate range.

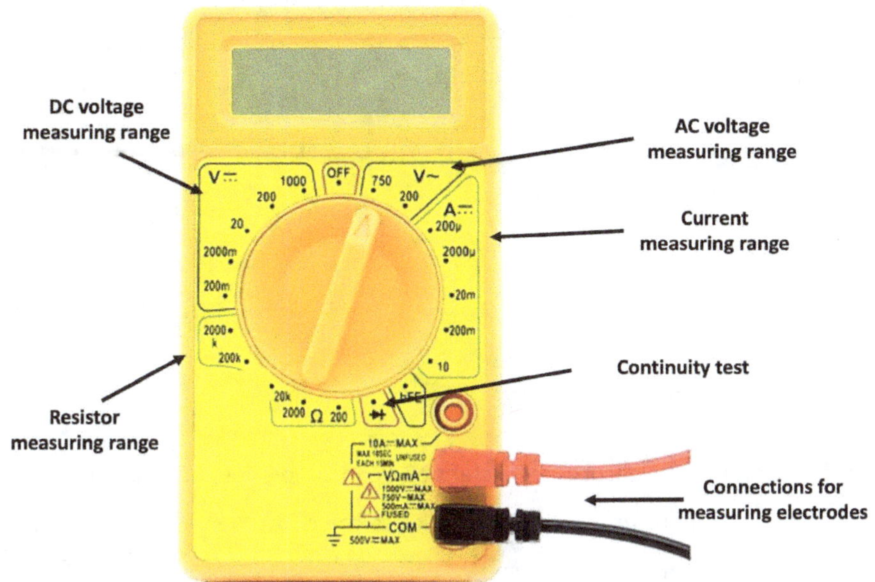

When making a measurement, always start from the highest possible voltage or ampere rating, or resistance value, and then turn the display setting down until a suitable value is displayed. This means, for example, that you turn the setting range to 200 volts if you make a measurement on a DC voltage source and suspect a value between 20 and 200V.

If you want to measure a voltage, you have to connect the measuring electrodes parallel to the voltage source or to the component you would like to measure. In the case of a light bulb, for example, this would work like this:

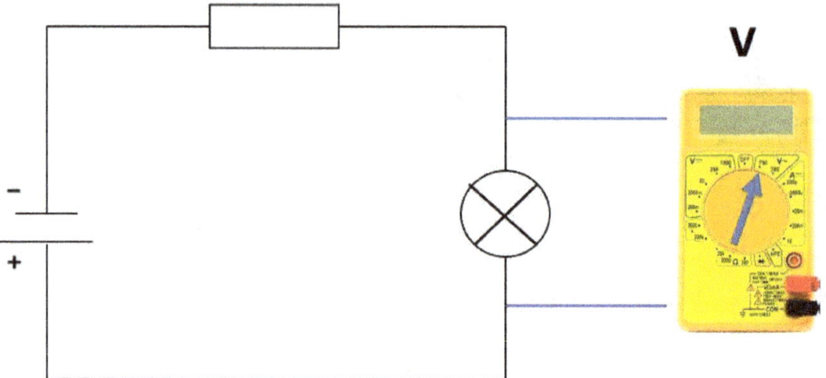

And if you intend to measure the current of a consumer, you have to connect the measuring instrument (multimeter) in series to the consumer, i.e. disconnect the line. This would then work like this:

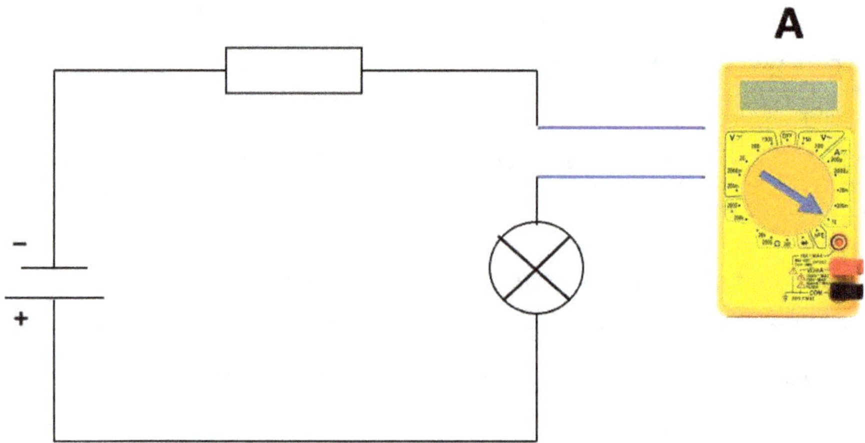

2.1.8 Series connection and parallel connection

Basically, you can connect electrical components in series (serially) or parallel to each other, or create a mixture of series and parallel connection. In the following, let's imagine several batteries that we connect either in parallel or in series and see what happens to the two parameters voltage and current.

14

<u>Series connection of the batteries:</u>

If the batteries are connected in series (connect one "-" and "+" pole of each two batteries), the voltage of the circuit increases, but the current remains the same. The resulting total voltage is obtained by adding the individual voltages. This means in this example with four 12V 110Ah batteries: U_{res} = 12V + 12V + 12V = 48V.

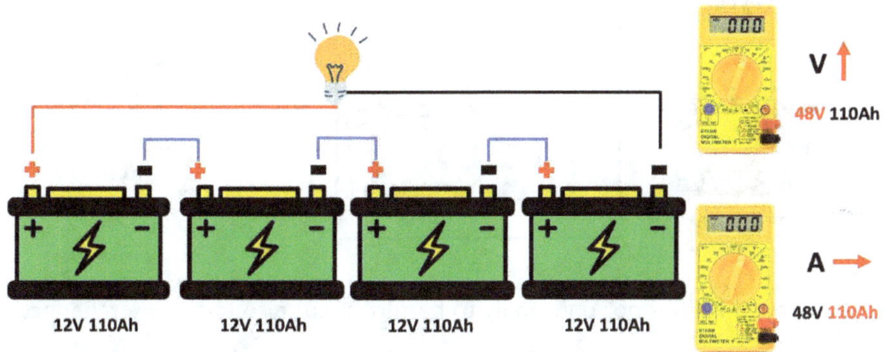

<u>Parallel connection of the batteries:</u>

When the batteries are connected in parallel (connect all "+" poles or all "-" poles to each other respectively), the current strength of the circuit increases, but the voltage remains the same, so the effect is exactly the opposite of the series connection. The resulting total amperage is obtained by adding the individual amperage. This means in this example with four 12V 110Ah batteries: I_{res} = 110Ah + 110Ah + 110Ah = 440Ah.

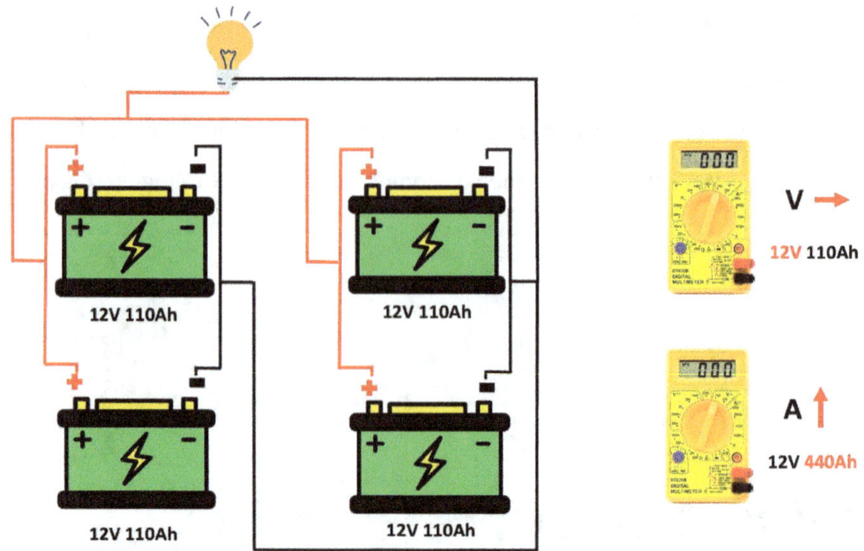

By the way, this does not only apply to batteries, but also to the interconnection of PV modules, for example. However, we will look at this in detail in a later chapter.

2.2 Fundamentals of semiconductors

When dealing with photovoltaics, one cannot avoid the term semiconductor. This is because the layer (photoactive layer) of a solar cell, which is crucial for electricity production, consists of a semiconductor material. In this section, we will take a simplified and brief look at the basics of semiconductors.

To understand the basics about semiconductors, let's first take a look at the structure of an atom, the smallest building block of a substance.

In the course of time, there were a variety of atomic models (e.g., spherical particle model, Rutherford's atomic model, Bohr's atomic model ...). An atomic model serves as a pictorial idea of what an atom looks like. In modern quantum mechanics, the so-called orbital model is used today, but it is difficult to handle for explanatory purposes. A popular and widely used model to describe the structure of an atom is the shell model, which is based on Bohr's atomic model. This shall suffice for our purposes here.

An atom has a charged nucleus in the center, which consists of neutrons and protons. Neutrons are neutral and have no charge, whereas protons are positively charged. In this way, the overall charge of the nucleus is positive. The atom as a whole is again a neutral particle. For this to be the case, an atom still needs negatively charged electrons as compensation. An atom has the same number of electrons and protons. Moreover, the charge of the electron (-) is equal to the opposite charge of the proton (+). Therefore, the charges balance towards neutrality.

The previously mentioned multitude of atomic models now mainly differs in where the electrons are located and how they move. The shell model says that the electrons are located on shells around the atomic nucleus and move on them. So, the electrons cannot move freely in space, but only on fixed orbits. One can imagine this movement restriction with a train, which can drive only on rails and must follow these. Each shell has a different energy level. The electrons that are in the outermost shell are also called **valence electrons**. A movement of these valence electrons means a flow of electric current.

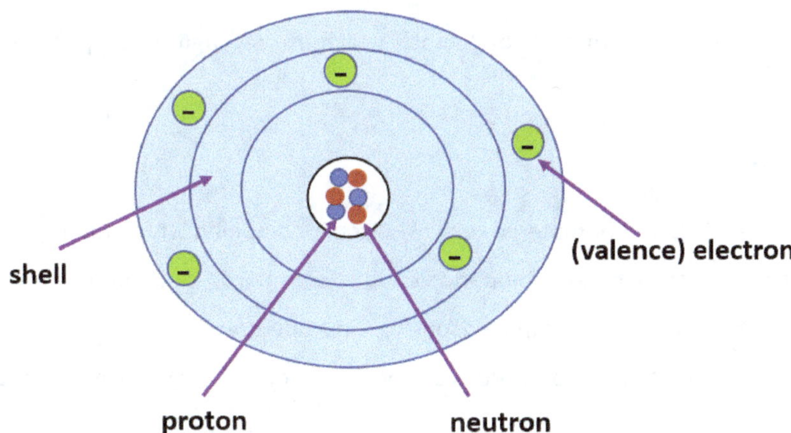

The following basics may go a bit deeper into detail and may be a bit more difficult to understand when reading for the first time and depending on previous knowledge. Nevertheless, it is useful to have heard the important terms in connection with semiconductors at least once. So stick with it as best you can!

In simple terms, **semiconductors** are materials whose electrical conductivity is lower than that of conductors (e.g., metal) and higher than that of insulators (e.g., plastic). Hence, the name semiconductor (half conductor). Silicon (Si) is the most common element used in the manufacture of semiconductor devices. The electrical properties of a semiconductor vary depending on the environmental conditions. A crucial property of semiconductors is their increased electrical conductivity when exposed to energy (solar energy in photovoltaic cells). By irradiation with sunlight, electrons are released (these are normally in fixed places) and are available for electrical conduction. This results in the formation of so-called holes. We will have a closer look at this in a moment. In contrast to conductors, the electrical conductivity of semiconductors also increases with increasing temperature (so-called hot conductors). To increase the conductivity of semiconductors enormously (factor 1,000,000), we have to add impurities (another element or foreign atoms). This process of impurity, i.e., the addition of foreign atoms to semiconductors, is called **doping.**

There are two types of semiconductors, which are distinguished by the type of doping.

P doping:

When a semiconductor material such as silicon is combined with another element of the third main group of the periodic table (e.g., boron or indium), a so-called **electron hole** is created. Simplified, one can imagine that an electron is missing at this hole site and that the properties of a positive mobile charge carrier are present here instead (but here only virtually – in contrast to the electrons). How does this work?

The element silicon has four electrons (**valence electrons**) in its outermost shell. Several silicon atoms together form a lattice-shaped crystal structure with four valence electrons each (i.e., a total of eight electrons each).

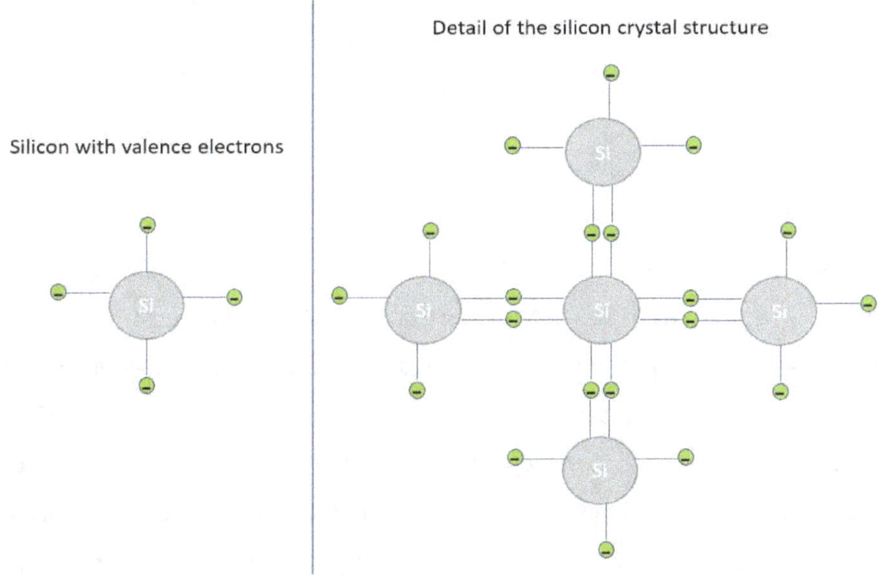

Boron, on the other hand, has only three electrons (valence electrons) in its outermost shell. If, for example, silicon is doped with boron, three electrons of the boron occupy three free electron positions in the outermost shell of the silicon, but actually four electrons are needed, i.e., one position remains unoccupied. This vacant position is the electron hole mentioned earlier. With high probability, electrons from the surrounding area just fill up this hole space (**recombination**). This creates an electron hole elsewhere. By the way, boron is called an **acceptor** because it accepts an electron. Note: In a **p-doping,** the positive electron holes predominate in the form of charge carriers.

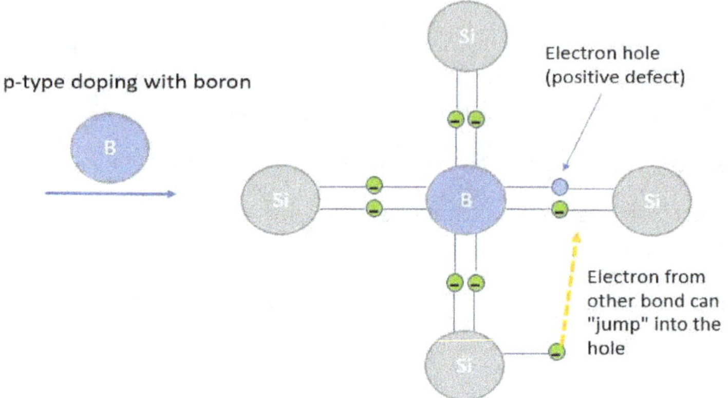

N doping:

However, when silicon is combined with an element of the fifth main group of the periodic table, which have five electrons in their outermost shell (e.g., phosphorus), we obtain an n-type semiconductor. In n-type semiconductors, one electron remains from the impurity element (e.g., phosphorus). Four of the electrons from the phosphorus occupy vacant sites in the outermost shell of the silicon, but one electron remains free to move. Phosphorus, by the way, is called a **donor** here because it supplies one electron. Note: In the case of **n-doping,** the negative electrons predominate in the form of charge carriers.

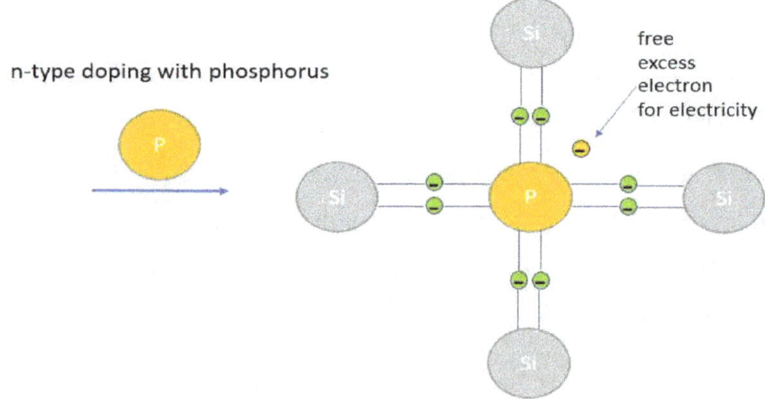

2.3 Fundamentals of photovoltaics

What is photovoltaics, and how is the electricity generated in a PV system? We would like to get to the bottom of this question in this chapter. The term photovoltaic is composed of the term "photos" (Greek) for light and the term "voltaic". The term "voltaic" stands for the eponym of the unit of voltage (V) Alessandro Volta (Italian physicist) or for the unit itself.

The basic principle of photovoltaics is the photoelectric effect. Albert Einstein took up this effect in 1905 and interpreted and explained it with the concept of energy quanta. Simplified, this effect can be explained in relation to a photovoltaic system as follows: Our sunlight consists of photons which, when they hit a solar cell, release electrons from its semiconductor surface and thus, through the flow of electrons, generate electricity. When a photon hits an electron, the electron moves, leaving a hole (electron vacancy). As we know, this electron flow is equivalent to a current flow.

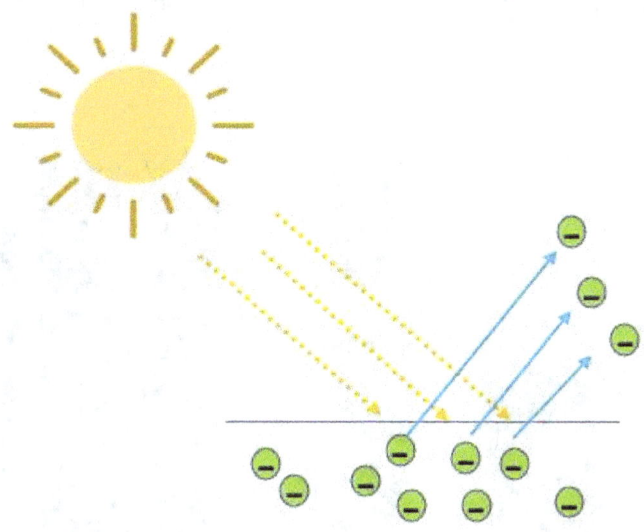

Every electromagnetic radiation (light is also an electromagnetic radiation) is quantized in photons. Photons have different energy levels and different wavelengths. Photons with shorter wavelengths have higher energy. The photons give up their energy to the electrons on impact.

The basic structure of a PV cell is a p-n junction, which generates direct current by absorbing solar radiation. We'll look at this in more detail in a moment. Since the resulting voltage and power of a single cell is low, several cells are connected in series and parallel to form a complete solar module, also known as a PV module.

It is important to mention at this point that solar modules foremost only generate direct current, but our household appliances need alternating current. To convert the direct current into alternating current, one or more inverters are used, depending on the size of the system. But more about that later.

2.4 Structure of a PV system, a PV module and a PV cell

A PV system consists of a large number of PV cells, each of which is connected together to form a PV module. These PV modules are in turn connected in series or parallel to form so-called PV strings and PV arrays, which in turn form the final PV system.

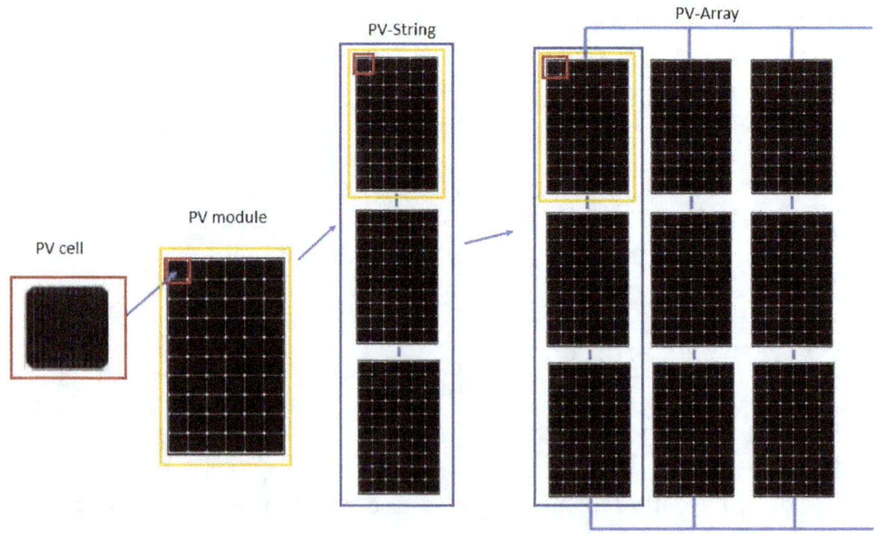

Before we look at the series or parallel connection of PV modules to form strings and arrays later on, we will first look at the structure of the smallest unit, the PV cell. As we already know, solar cells consist of doped semiconductor materials. The manufacturing process – in the case of monocrystalline silicon cells – is as follows: In the first step, quartz sand (SiO_2) is converted into raw silicon. This process takes place at very high temperatures. From this raw material, purest silicon is then extracted in further steps.

Further, monocrystalline silicon is produced from poly crystalline silicon using the so-called Czochralski process. In this step, the silicon is also doped with foreign atoms (e.g., boron and phosphorus) to obtain n-type material and p-type material. In the final step, the silicon is cut into very thin circular plates, known as "Wafer".

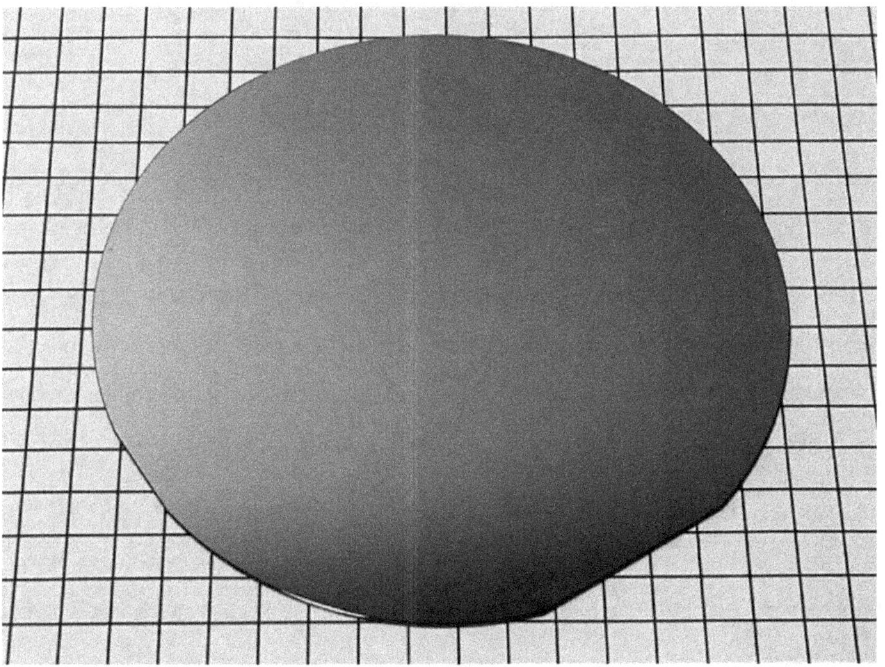

These wafers are doped differently on both sides. A p-n-doped solar cell can then be produced from them.

PV cells usually consist of very thin "Wafers" with a thickness of only about 0.1 mm. With the previous considerations, we can now look at the schematic structure of a PV cell. To do this, we bring together one p-doped and one n-doped material so that we obtain a p-n junction. The schematic structure of a PV cell is described in the following figure:

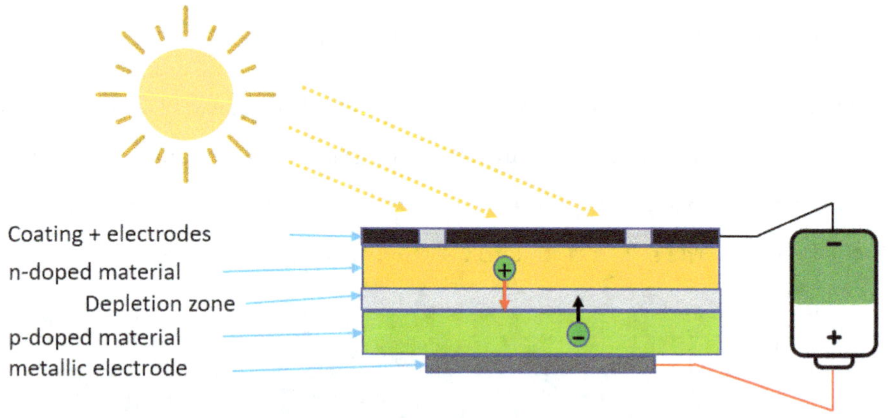

If we remember the behavior of the p-doped as well as the n-doped material from chapter 2.2, we can now see how the current flow in a solar cell is generated.

The excess electrons of the n-layer (electron surplus) fill the holes of the p-layer (electron deficiency) at the contact point between the two layers. In this area (contact area between n-layer and p-layer), a so-called boundary layer (also: depletion zone, space charge zone, junction layer) is formed.

As a result of this process, electrons are now missing in the n-layer, since they have migrated to the p-layer as described above. The n-layer is now positively charged (electron deficiency). The p-layer, on the other hand, is negatively charged (excess electrons).

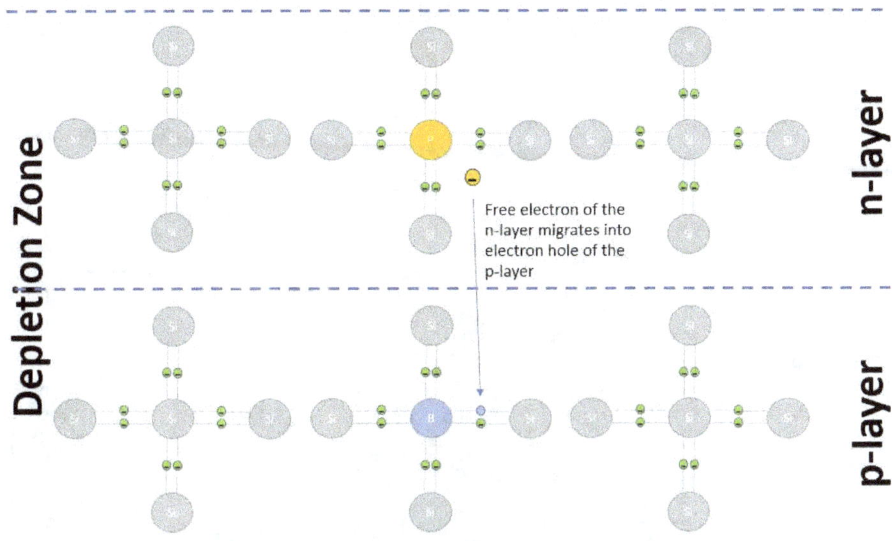

Now we let the sun shine on the boundary layer of the solar cell. As we have already learned from the photoelectric effect, the photons of the solar radiation can dissolve the electrons in the boundary layer again and both holes and free electrons are created. The freed electrons are attracted by the n-layer and move upwards, where they can be picked up by a metallic contact (electrode). The positively charged holes, on the other hand, appear to migrate to the negatively charged p-layer (no actual movement; only virtual imagination). Again, there is a contact that closes the circuit with a load. This process creates the current flow (electron migration: electrons in motion = current flow). During this migration, however, it also happens in some places that the free electrons, fill the holes (recombination; see chapter 2.2) and are thus no longer available for the current flow. Therefore, one tries to keep the recombination as low as possible.

By the way, the n-layer is much thinner than the p-layer and therefore translucent. This is what allows the photons to reach the boundary layer in the first place. The electron migration (incl. recombination) is shown again schematically in the following figure:

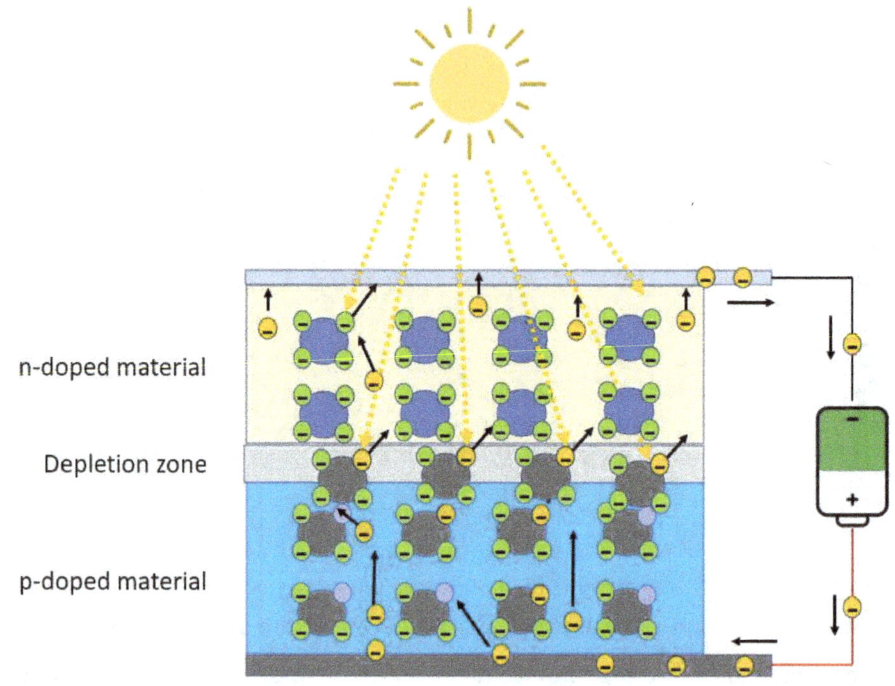

Most solar modules consist of 60 or 72 cells. However, there are also modules with 36 cells or even with 96 cells. On the upper side, there is usually an anti-reflective coating, which gives the solar cell its characteristic dark color.

The electrical contacts that connect the solar modules are called "Busbars" and "Fingers". The screen printing process is used to print these contacts (usually made of silver) on the sun-facing side of the solar cell. These contacts serve to pick up or pass on the electrons that begin to flow as a result of the process described above. The "Finger" collect the direct current and pass it on to the "Busbars". On the bottom side of the solar cell, there is a simple electrode as a counter pole. Thus, a consumer can be connected, and the current flow is guaranteed.

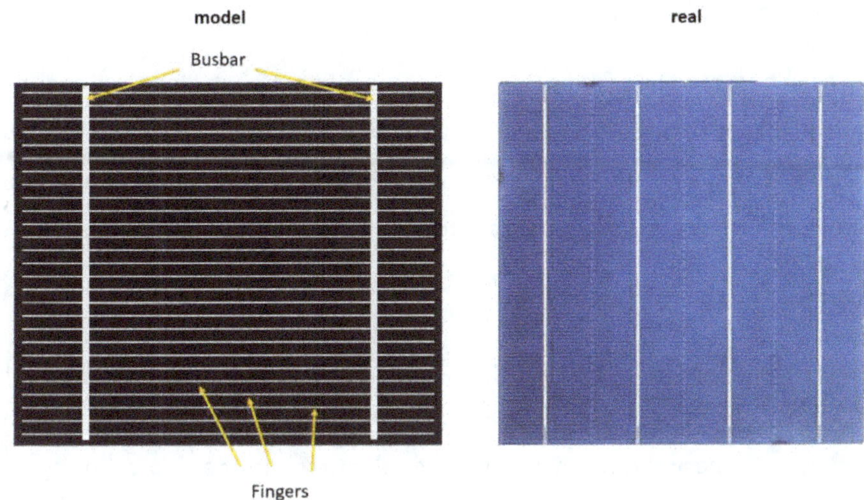

In the following, we will also take a look at the structure of a PV module, which – as we already know – consists of several individual PV cells. In order for the module to function optimally and also be protected against the effects of weather, a few more components are necessary.

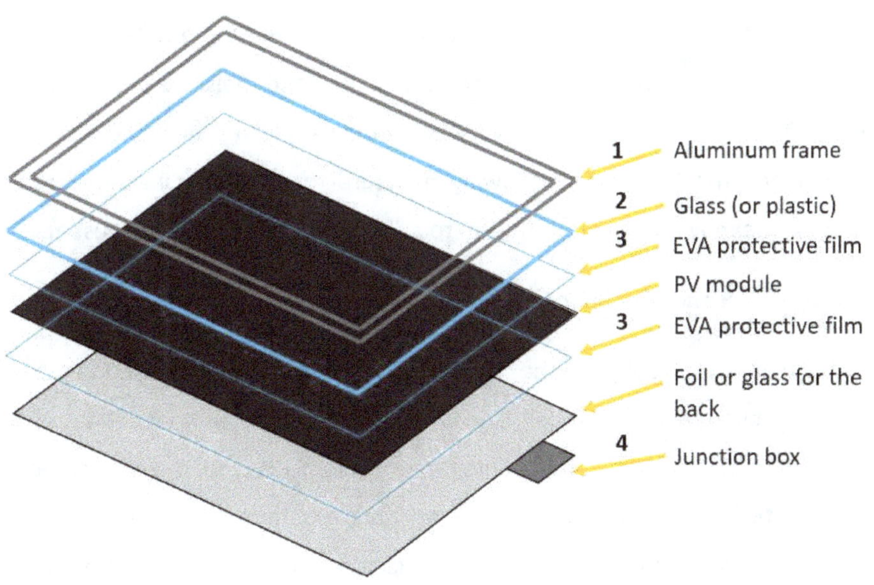

1) Frame made of aluminum:

The aluminum structure forms the basis for mounting the solar modules and their components. It also provides a connection point for grounding so that any fault currents can be safely conducted into the ground. Aluminum frames are lightweight and yet relatively resistant to mechanical stress.

2) Glass:

Solar modules have glass on the front side. The glass of the solar modules is usually coated with an anti-reflective coating to ensure maximum absorption of sunlight. In addition to the anti-reflective coating, the glass also provides protection against environmental factors. The glass material of solar modules is strong enough to withstand mechanical stress. Even hail is typically not a problem. Alternatively, a plastic with similar properties can be used here.

3) EVA protective film:

EVA is the abbreviation for ethylene vinyl acetate. The purpose of this protective layer is to encapsulate the photovoltaic cells on both the top and bottom sides. EVA encapsulation provides protection against dust and moisture penetration. If the quality of the EVA layer is not very high or is damaged over time, this can in turn damage the PV cells by allowing moisture to penetrate inside the solar modules. The moisture can in turn lead to a short circuit between the busbars.

4) Junction box and connections:

The junction box is located on the back of the solar module. This is used to tap the current supplied by the solar cell. The junction box can be equipped with diodes (bypass diodes). If a cell of the module does not supply current due to a defect or contamination, the remaining current can take a detour, so to speak, via this bypass diode. You can think of it like car traffic (electrons can be thought of as cars, the wires as roads). Junction boxes are usually equipped with MC4 connectors.

MC4 plugs are weatherproof plugs that have two pins, a male pin for the positive terminal and a female socket for the negative terminal.

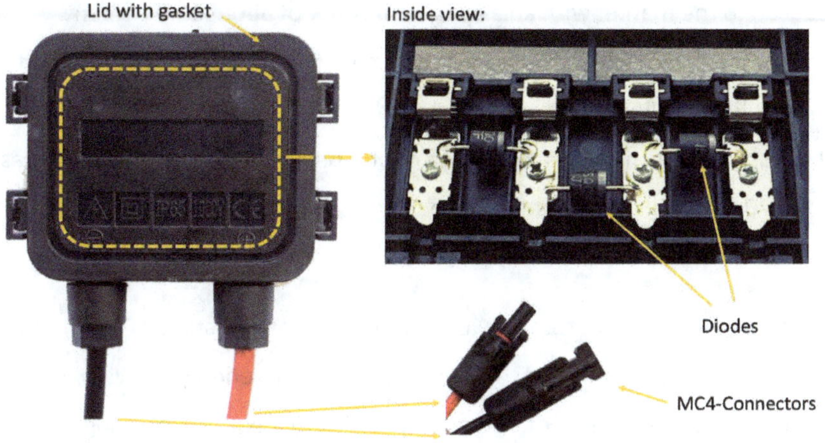

3 A brief history of photovoltaic development

3.1 The different types of PV modules

The use of solar energy to generate electricity began in the 20th century, more precisely in 1954, with the presentation of the first solar cell. This was developed in the "Bell Laboratories" research laboratory, in the USA.

The first solar cells had a relatively low efficiency and were also very expensive. One of the first applications was in space travel to generate electricity using sunlight. In the late 1970s, solar cells were used in remote areas of the United States to generate electricity where there was no power grid. Even at that time, the cost of solar cells was still very high and few people could afford the technology. However, as technology has advanced, the cost of solar panels has dropped significantly over time.

In 1980, the company "Arco Solar" was one of the first companies to manufacture PV panels. The company commissioned a 1 MW (1,000,000 watt) solar farm in California (USA) in 1982. Then, in 1986, the first thin-film solar module was introduced to the market with an average efficiency of about 15.9 percent. We will find out what thin-film modules are in a moment. In 2000, the total capacity of solar modules installed worldwide reached the 1 GW (1,000,000,000 watts) mark for the first time. The market share of the photovoltaic industry gradually increased and in 2020 the total global installed capacity of photovoltaic systems reached 760 GW.

It can be assumed that this total power and the number of installed PV systems will still increase strongly in this and the next decades due to energy and climate crises. The sun is the largest source of energy in our solar system, it is only logical to use it to meet our energy needs.

Solar technology has developed steadily since its beginnings and different types of PV modules have been developed. We have looked at the schematic structure of a solar cell and the formation of the current flow in a PV cell using the example of a monocrystalline solar module. However, there is now a wide range of different modules and technologies.

In addition to monocrystalline cells, there are also poly crystalline cells, for example. Monocrystalline and poly crystalline modules differ in their efficiency. The efficiency of monocrystalline modules (approx. 20-25%) is higher than that of poly crystalline modules (approx. 15-20%), but the costs of poly crystalline modules are lower. Poly crystalline solar modules are also well suited in areas with high ambient temperatures, so poly crystalline solar modules are usually used in very warm areas. Depending on the efficiency (e.g., high efficiency = less area or modules for the same power), cost of the modules (monocrystalline modules are more expensive than poly crystalline) and other factors, a decision can be made for poly crystalline or monocrystalline modules.

30

In the beginning, these two types of modules were the only ones to choose from. Of course, technological development has not come to a standstill in the photovoltaic industry, and there are now several other types of PV cells. We are now in the third generation of PV cell development.

1st generation	2nd generation „Thin film"	3rd generation „New Technology"
Monocrystalline cells	Amorphous silicon thin film cells	Nanocrystal-based cells
Polycrystalline cells	Cadmium telluride (CdTe) Thin film cells	Polymer-based cells ...

Currently, the most widely used types of solar modules are monocrystalline PV modules, poly crystalline solar modules and thin-film solar modules. We will take a closer look at these in the following.

3.1.1 Monocrystalline PV modules

Monocrystalline solar modules are manufactured from pure silicon using the so-called Czochralski process. In this process, a so-called seedling (seed crystal) is placed in a bowl of molten silicon and a large cylindrical crystal, called "Ingot", is pulled from it. You can think of it as wax candle pulling, if you've ever done that. The best way is to watch a video about it. From the "Ingot" the thin "Wafer" are then cut again. The monocrystalline cells are then cut into a square shape. The corners are beveled or removed. The cells have a dark blue to black color. The advantages of monocrystalline cells are: a) high efficiency (20-25%), b) long life (at least 20 years), c) robustness, d) high electricity yield with a small roof area. The disadvantages of monocrystalline cells, on the other hand, are: a) expensive to purchase, b) poor environmental performance in production, c) sensitive to high temperatures.

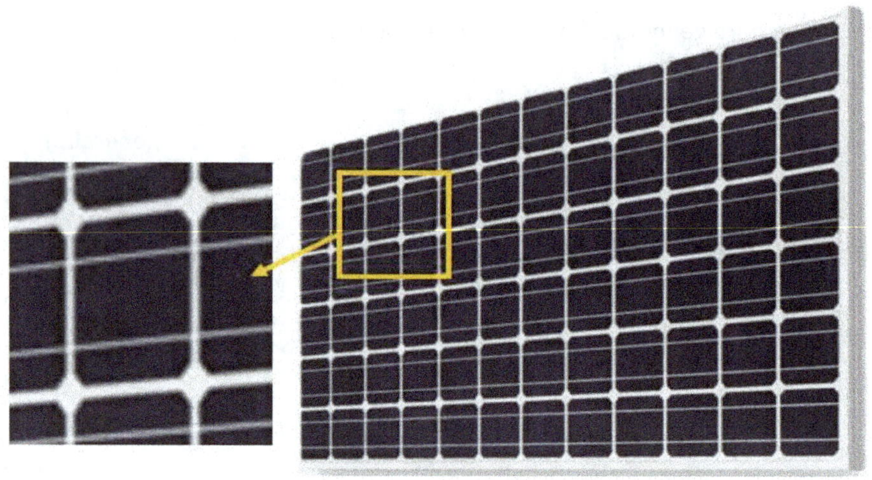

3.1.2 Poly crystalline PV modules

Poly crystalline solar modules are the newer form of solar modules compared to monocrystalline solar modules. They have good resistance to harsh environmental conditions. Poly crystalline solar modules are also made of silicon and are manufactured by assembling silicon fragments that give the PV module its characteristic appearance. The assembled silicon fragments are then also cut into thin "Wafer".

Poly crystalline solar cells usually have a bluish color and are clearly recognizable by their appearance. The shape of poly crystalline solar cells is rectangular, and therefore there is less waste in the manufacturing process. The advantages of polycrystalline cells are: a) insensitive to high temperatures, b) cheaper than monocrystalline cells, c) easier to manufacture and better environmental performance than monocrystalline cells. On the other hand, the disadvantages of poly crystalline cells are: a) lower efficiency (15-20%), b) more surface area for the same electricity yield (compared to monocrystalline).

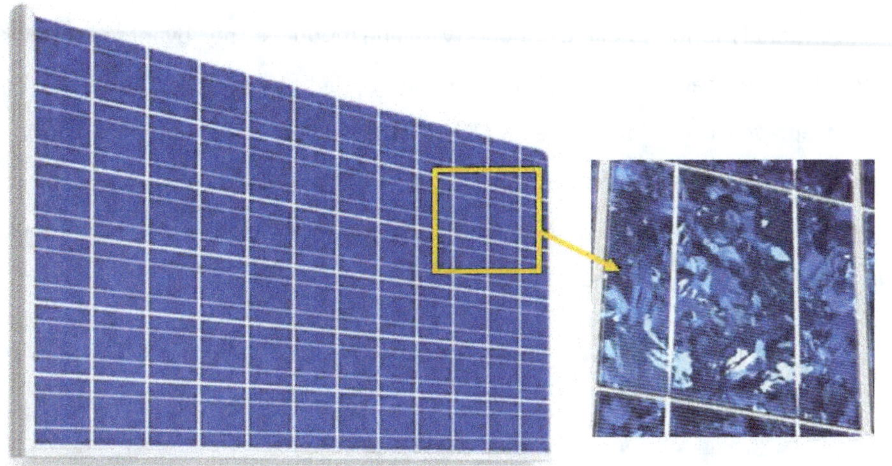

If there is enough area for the desired PV power, you can definitely choose the cheaper and more environmentally friendly poly crystalline variant. Unless the appearance is strictly rejected for aesthetic reasons or the system is to provide the maximum power for the available area, then the choice must fall on monocrystalline modules. But before deciding, one should also consider the thin-film modules.

3.1.3 Thin-film solar modules

Apart from polymer-based and nanocrystal-based third-generation solar cells, which are not considered here, second-generation thin-film solar cells are the most advanced development in the photovoltaic industry. Unlike the production of poly crystalline and monocrystalline solar cells, thin-film solar cells are not always made of silicon. A thin-film cell can also be made of cadmium telluride (CdTe) or copper indium gallium selenide (CIGS), for example. However, silicon can also be used, but then in an amorphous structure. The term amorphous describes the structure of the silicon, which in this case is not crystalline (ordered lattice structure) but has a disordered atomic structure (amorphous). The materials are vapor-deposited in a thin layer onto a carrier film during the production of the PV cell. Hence, the name thin film PV cell. This type of solar cell can be hundreds of times thinner than conventional monocrystalline or poly crystalline modules. The advantages of these modules are mainly a) cheap production, b) light weight, c) flexibility. The disadvantages are: a) low efficiency (10-15%) b) high area consumption, c) difficult installation (because frameless and thin). The high space consumption due to the low efficiency is one of the main reasons why these modules are rather not found in the private sector on house roofs.

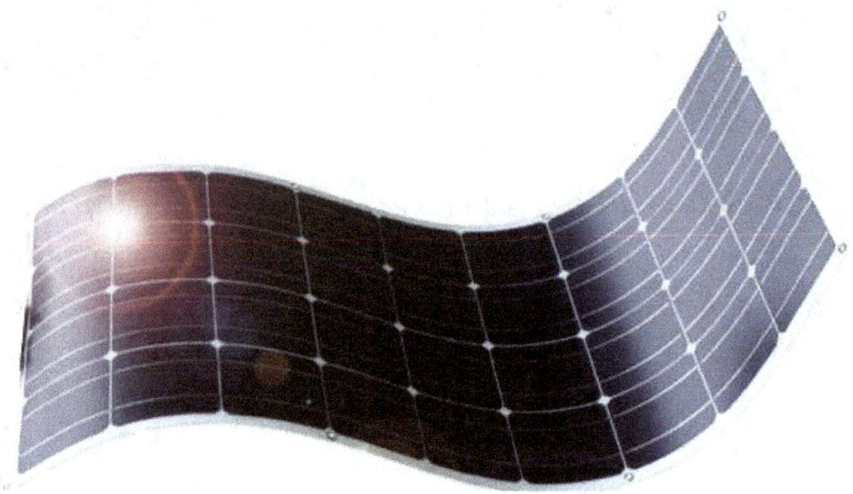

3.2 Important key figures in connection with PV modules

If you inform yourself about PV modules or PV systems for the first time, you will be confronted with many key figures and data in the data sheets. For example, you may have wondered what the Wp or W_{peak} or "Watt peak" means, or what to do with the open circuit voltage? Let's take a look at the most important electrical data for PV modules.

Each PV module has its own current-voltage characteristic (I-V characteristic or I-V characteristic). The I-V characteristic curve of a solar cell describes how much solar energy the photovoltaic cell can convert into usable electricity. That is, the I-V characteristic curve provides information about the efficiency or maximum power that a single module, string, or array can deliver. One can create an I-V curve for a single module, as well as for an entire PV array. The power that a cell can deliver to the load is the product of voltage and current ($P = U \times I$).

In such a current-voltage characteristic curve, this curve between current and voltage is shown, e.g., as a function of temperature (e.g., 50 °C in the figure) or irradiance (e.g., 1000 W/m² in the figure). The peak value for the power of the solar module, which is called the maximum power point or "maximum power point" (MMP), can be read here. This point of maximum power of a PV cell depends, among other things, on the temperature of the cell and the irradiance and is therefore not a fixed point, but variable. At this point of maximum power, short-circuit current and open-circuit voltage have their optimum values. These two values are relevant for the design of a PV system. Short-circuit current and open-circuit voltage can also be read from the I-V characteristic. The short-circuit current I_{SC} is simply the current that flows when the voltage is (close to) 0V, i.e., when the two poles of the PV module – without a load in between – would be connected to each other, i.e. short-circuited. The open-circuit voltage V_{OC} or U_{OC}, on the other hand, is the voltage value at which no consumer draws current (0A). A PV system must be designed according to the lowest expected temperature (relevant for open-circuit voltage) of the PV modules and according to the highest expected irradiance (relevant for short-circuit current).

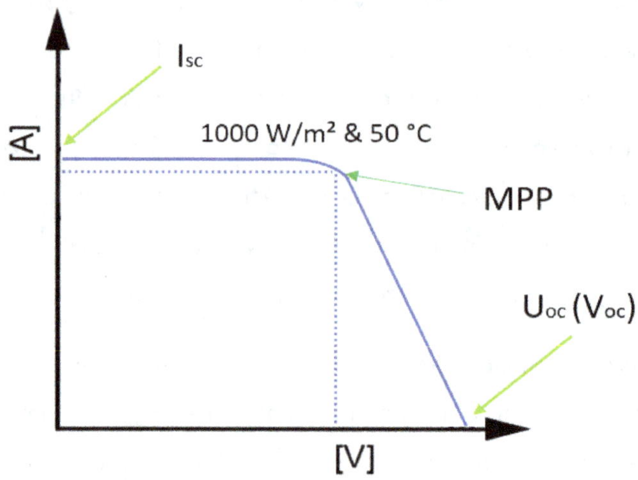

The standard conditions (STC) for maximum efficiency of a solar module are a temperature of 25 degrees Celsius, an irradiance of 1000 W/m² and an air mass of 1.5. These conditions are commonly referred to as STC conditions. Each module has its degradation coefficient, the efficiency of the module decreases as the temperature increases.

Other important terms related to photovoltaic modules are fill factor ("Fill factor") and performance ratio ("Performance Ratio").

The fill factor (FF) is calculated with the following formula: $FF = P_{MPP} / (U_{OC} \times I_{SC})$ and describes the maximum power of a PV cell related to its open circuit voltage and short circuit current. The fill factor thus relates the nominal power at the "maximal power point" to the open-circuit voltage and short-circuit current. It can be considered as the quality factor of a PV cell. The higher the fill factor (usually values of 0.5 - 1; 1 would be the theoretical ideal) the higher the quality of the PV module.

The "Performance Ratio (PR)", on the other hand, indicates the ratio between possible and achieved yield and is given as a percentage. The achieved yield can be easily read off the PV meter. The closer the ratio is to 100%, the more effective the system is.

36

If you read up on individual PV modules, you will come across even more terms, such as Wp. What does Wp mean? Wp or W_{peak} stands for "Watt peak" and indicates the maximum power – the nominal power – of a solar module when it is operated under STC conditions (cell temperature = 25 °C, irradiance = 1000 W/m², air mass "AM" = 1.5). PV modules can also be compared with each other using this value. In practice, however, the Wp values specified in each case are usually not achieved, since the environmental conditions vary. kWp in this context, by the way, then simply stands for 1/1000 Wp. This means, for example, that 550 Wp is 0.55 kWp.

Rated power: Designated as P_{max} or P_{nenn} or P_{MPP} [W]. Indicates the maximum power under standardized conditions. Often given in Wp or kWp. The specification is often higher than is achieved in actual operation.

Nominal voltage: Mostly referred to as V_{MPP} or U_{MPP} [V]. Indicates the voltage at the "maximum power point" for the current irradiation.

Rated current: Generally denoted by I_{MPP} [A]. Indicates the current at the "maximum power point" for the current irradiation.

Open-circuit voltage: The voltage applied to the PV system when no load is connected, typically referred to as V_{OC} or U_{OC} [V]. Measurable with a multimeter.

Short-circuit current: The maximum current of the PV system that flows if there is no load between the poles, but they would be short-circuited. Generally referred to as I_{SC} [A].

Efficiency: Denoted by η [%]. The efficiency indicates how much incident energy (solar radiation) the PV module can convert into electricity. The higher this value, the better or more efficient a PV module works, i.e. it can generate more electricity than another module under the same conditions. The efficiency is determined under standard test conditions (STC).

Irradiance: Irradiance [W/m²] is the power of electromagnetic radiation incident per m² on a surface (e.g., the PV module).

Standard Test Conditions (STC): Test environmental conditions for the PV module. 1000 W/m² at 25 °C and air mass "AM" 1.5.

Nominal cell operating conditions (NOCT or NMOT): Test environmental conditions that should be closer to normal operation (natural environmental conditions) in order to obtain more meaningful values. E.g., 800 W/m² at 20 °C and air mass "AM" 1.5 as well as wind 1-2 m/s and cell temperature of 40-50 °C – depending on the specification.

Example from a data sheet of a PV module:

ELECTRICAL SPECIFICATIONS			
STC rated output (P_{mpp})*	300 Wp	305 Wp	310 Wp
PTC rated output (P_{mpp})**	273.2 Wp	277.9 Wp	282.5 Wp
Standard sorted output			0/+5 Wp
Warranted power output STC ($P_{nominal}$)	300 Wp	305 Wp	310 Wp
Rated voltage (V_{mpp}) at STC	35.74 V	35.77 V	35.80 V
Rated current (I_{mpp}) at STC	8.40 A	8.53 A	8.68 A
Open circuit voltage (V_{oc}) at STC	45.16 V	45.29 V	45.42 V
Short circuit current (I_{sc}) at STC	8.91 A	8.95 A	8.99 A
Module efficiency	15.5%	15.8%	16.0%
Rated output (P_{mpp}) at NOCT	209.5 Wp	213.0 Wp	216.5 Wp
Rated voltage (V_{mpp}) at NOCT	32.63 V	32.67 V	32.70 V
Rated current (I_{mpp}) at NOCT	6.42 A	6.52 A	6.62 A
Open circuit voltage (V_{oc}) at NOCT	41.44 V	41.56 V	41.68 V
Short circuit current (I_{sc}) at NOCT	6.89 A	6.92 A	6.95 A

4 PV systems and their components

By now, we know how a PV cell works and how different PV modules are constructed and manufactured from PV cells. In this chapter, we will take a look at the other components of a PV system. Besides the PV modules, we need a few more components to make our PV system work. Here, we can mainly distinguish between two systems in terms of structure and the required components of a PV system. Depending on the application (power feed-in, own use, mixed form), a grid-connected system (on-grid) or a stand-alone system (off-grid) can be considered. In this chapter, we will take a look at the structure, components, installation options and acceptance or commissioning of these systems.

4.1 On-grid vs. off-grid systems

Photovoltaic systems can be divided into two groups. The first group is grid-connected systems (on-grid systems). A grid-connected system is a PV system that is connected to the public power grid. On-grid systems can either be full feed-in, i.e., sell all the electricity to the grid operator, or only partial feed-in.

In the case of partial feed-in, the system can be designed, for example, so that one's own electricity consumption is covered as well as possible and only the surplus electricity (e.g., when the sun shines very long and intensively on a day) is fed into the public grid. The advantage of grid-connected systems is that, depending on demand, either electricity can be purchased from the grid operator (e.g., when the sun is not shining) or excess electricity can be fed into the public grid (e.g., when electricity consumption is low but the PV system supplies a lot of electricity). The following schematic diagram serves as a first orientation, how such a grid-connected system is roughly structured. The details will follow!

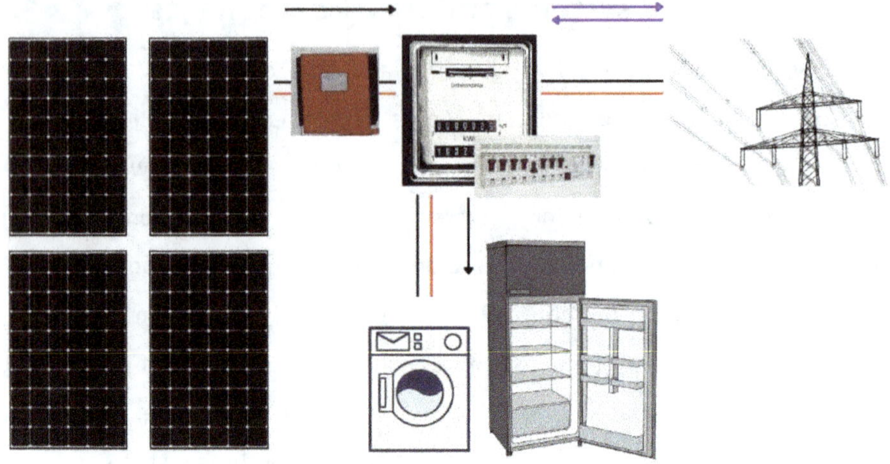

The second group consists of the so-called stand-alone systems (off-grid systems). These systems create a self-sufficient power supply and can be seen as an independent mini power plant.

These systems are mainly used in remote areas (e.g., developing countries, houseboats, mobile homes …) where there is no power supply from the public grid, but also in all areas of application (e.g., Tiny House) where maximum self-sufficiency is required. Since the sun does not always shine, but electricity is usually needed continuously (e.g., for the refrigerator), a power storage system (battery and/or heat storage) is needed for an off-grid system. These systems are also often coupled with an emergency generator or wind turbine to provide a continuous power supply. The following schematic illustration again serves as a first orientation for the construction of an off-grid system. In this figure, a black box is integrated, which stands for further components of the PV system. For the sake of clarity, we will look at these in detail in the following sections.

Both on-grid and off-grid systems can be planned with battery storage. The advantage of this is obvious: surplus electricity can be stored in times of high yield and/or low consumption to provide solar power even when the sun is not shining for a few days. Depending on how large the battery storage and PV system are designed, you can bridge more or less days without sunshine. This is always a question of cost and the individual location of the PV system.

Which system should you choose, and are there differences in the structure of the two systems? We will take care of this in the following. Of course, it depends on your individual situation, but unless you are living with a primitive people in the Amazon or on a houseboat, then a grid-connected system (on-grid) is probably more likely, at least for your own home. Then there is the question of whether you want to fully feed the electricity into the grid or consume it in your own household. With rising electricity prices and falling feed-in tariffs, it is now more worthwhile than ever to consume the electricity yourself and only feed in surplus electricity.

In the following, we will therefore deal with an on-grid system and its components (including battery storage). The schematic structure incl. connection is shown in the following figure. We need: **1)** solar modules, **2)** junction box & DC overvoltage protection, **3)** inverter, **4)** solar electricity meter, **5)** AC overvoltage protection, **6)** main distribution (electricity box) with electricity meter, **7)** house connection to the public electricity grid, **8)** equipotential bonding (grounding), **9)** consumer, **10)** optional: electricity storage.

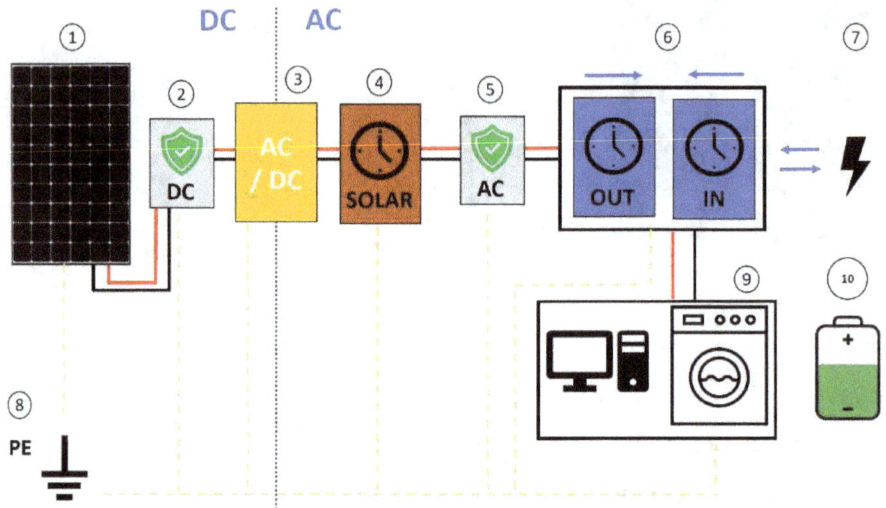

We will look at the individual components in detail in the next section. Before that, however, some information about the integration of an electricity storage system. A power storage system can be integrated into a PV system in two different ways. You can either plan the electricity storage as a DC coupled electricity storage (direct current system), or as an AC coupled electricity storage (alternating current system). Let's take a closer look at these two variants.

4.1.1 DC coupled power storage

A DC-coupled power storage unit (**10**) is connected to the DC side of the system, i.e., directly behind the solar modules (after junction box and DC overvoltage protection). For the connection – apart from the current storage – a DC-DC converter (**11**) and a charge controller (**12**) are also required. The DC-DC converter regulates the voltage coming from the PV modules down to an optimal charging voltage. The charge controller regulates the charge current and the charge voltage with which the electricity storage system is charged and has the task of charging or discharging the electricity storage system as efficiently as possible. An integrated deep discharge protection also prevents deep discharge and thus protects the power storage unit from a defect.

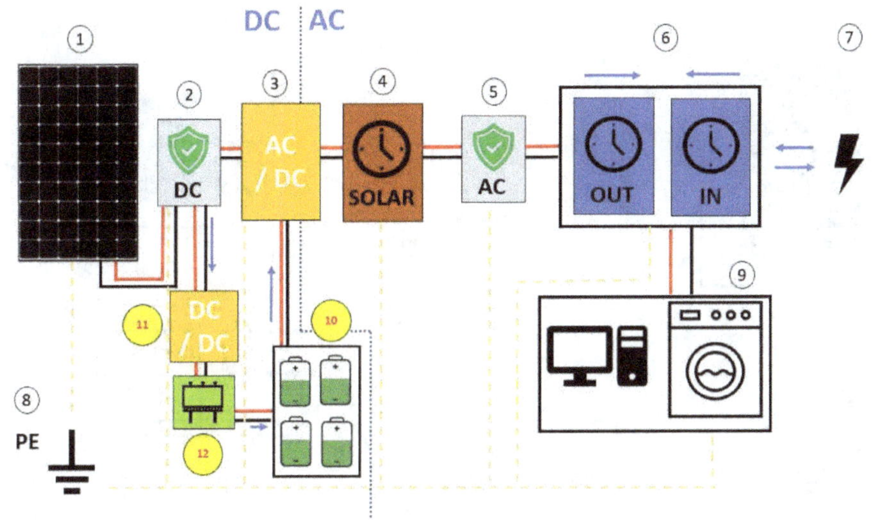

When electricity from the electricity storage system is needed (e.g., at night), it is fed into the AC grid of the house via the DC-AC inverter (**3**). DC coupled electricity storage systems are mostly used in simple and small PV systems. **A DC system is very interesting when planning a new system** because it has a high efficiency, does not require much space and is comparatively easy to install. **For retrofitting existing PV plants, the system is rather less suitable**.

4.1.2 AC-coupled power storage system

An AC-coupled electricity storage system, on the other hand, is connected to the AC grid, i.e., only after the PV inverter. A battery inverter (**11**) is required for this purpose, which converts the incoming AC current of the AC system into DC current or converts outgoing DC current from the charge controller (**12**) back into AC current. The power storage system (**10**) and the charge controller (**12**) can be identical to the DC coupled system. AC-coupled electricity storage systems are mainly used for larger PV systems and are particularly recommended **when retrofitting an electricity storage system for an existing PV system** (no need to replace the PV inverters).

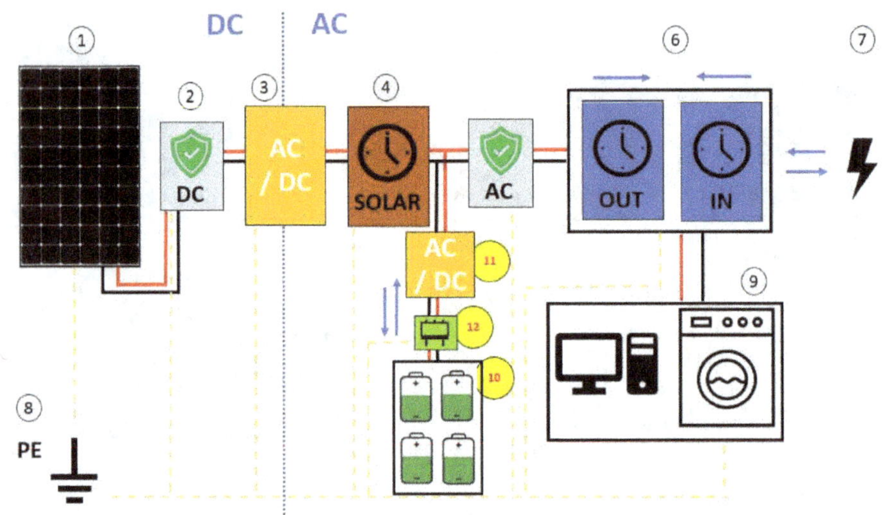

4.2 The components of a PV system in detail

In this chapter, we will now look in detail at the components of a grid-connected PV system with electricity storage. The **following numbers (1-12) after the respective chapter heading refer to the previous figures.**

4.2.1 Photovoltaic modules & solar cables (1)

The components that serve to generate electricity in a PV system are the PV modules, the structure of which we have already dealt with in detail. As we know, there are monocrystalline, poly crystalline and other types of PV modules. However, for the connection of these modules and for the other components, this distinction does not matter. All these modules produce direct current. Special solar cables (special DC cables) and connectors (usually MC4 connectors) are used to connect the PV modules together – starting from the junction boxes on their rear side – and to transmit the produced current. Cables and connectors should be laid in weather-protected installation ducts. The cross-section of the cables must be determined depending on the PV system. The PV modules must also be grounded.

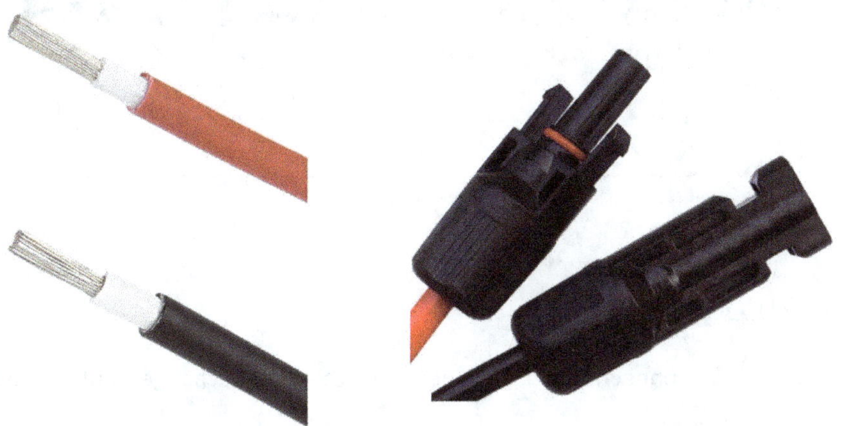

Depending on the PV module, you can expect a voltage between 30 - 50 V (DC) and a current of up to 5 A. You can connect several PV modules in two ways. Either with a series connection or with a parallel connection (or also combined).

<u>Series connection:</u>

In a series connection, you simply connect one pole of a solar module to the opposite pole of the next solar module ("+" with "-" or "-" with "+"). As we already know, in a series connection, the voltages of the individual components add up. The current does <u>not </u>change. For example, if you connect four solar modules together, each with 30V output voltage, you get 30V + 30V + 30V = 120V output voltage. The series connection of PV modules is then called a "String".

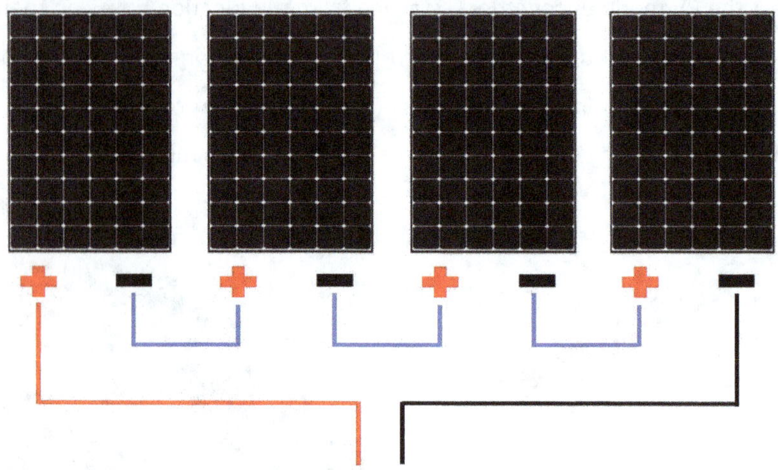

<u>Parallel connection:</u>

With a parallel connection, on the other hand, the current changes, but the voltage remains the same.

In order to add up the currents of the individual modules, the solar modules must be connected to each other as follows. Connect the same poles of the individual modules to each other ("+" with "+" or "-" with "-"). Assuming that one solar module supplies 5A current, the total current of all four modules is: 5A + 5A + 5A = 20A.

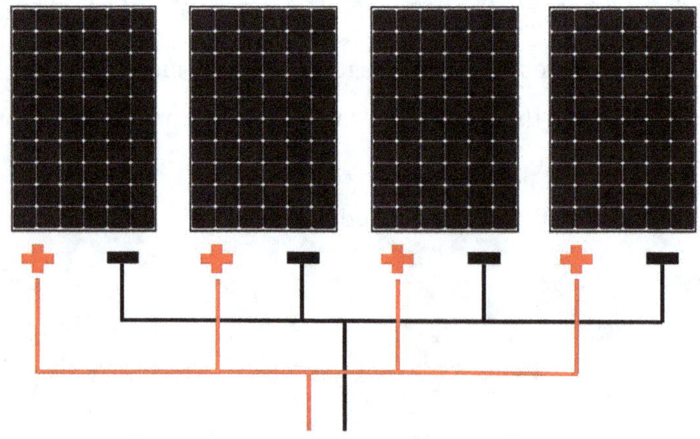

In most cases, a combination of series and parallel connection of the modules takes place to achieve the desired performance of the PV system, as well as to maintain voltage and current values. First, the "Strings" (series connection) are formed and then connected in parallel to an "Array".

4.2.2 Module junction box with DC overvoltage protection (2)

The current now comes from the solar system first into a junction box, which contains a surge protector. For every 10 m of line length – both on the direct current side (DC side) and on the alternating current side (AC side) – one surge protection device is required. Here, a differentiation is also made according to the number of "Strings" installed. The illustrated junction box with surge protection is suitable, for example, for connecting two PV "Strings" (it has one input and output per "String").

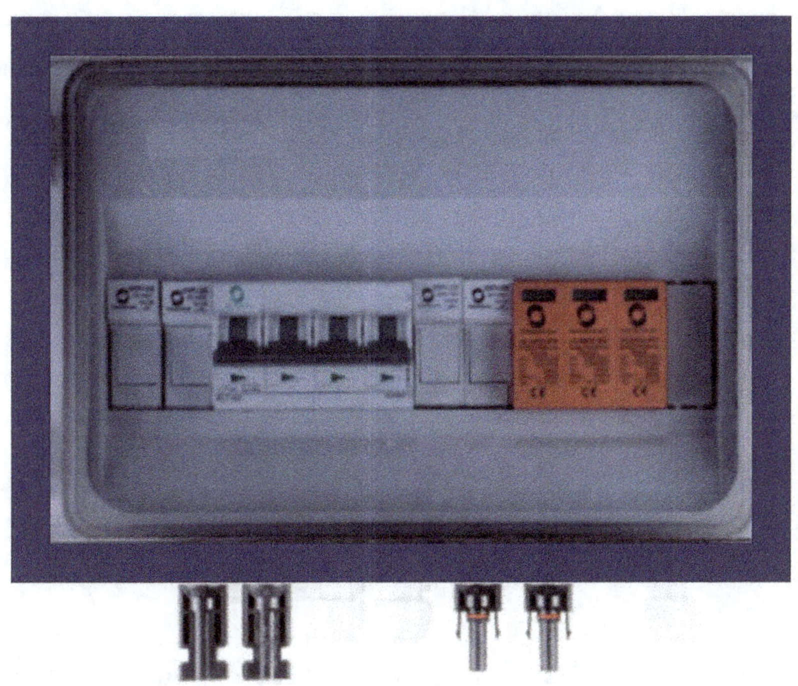

The overvoltage protection is needed as protection in case of lightning and other – also internal – overvoltages, so that these are safely discharged and cannot cause any material or immaterial damage.

4.2.3 DC-AC inverter (3)

<u>Note:</u> In this book we assume that our devices require 230V 50Hz mains voltage. If you have devices that require a different mains voltage (e.g., 110V or 120V 60Hz), then you must purchase components (e.g., inverters) that support this voltage.

From the DC overvoltage protection, the current then reaches the inverter via connection cables. This has the task of converting the direct current of the PV modules into an alternating current. An inverter converts the direct current into a sinusoidal alternating current with the help of semiconductor switching elements and pulse width modulation. However, we will not look at how exactly this works here. There are also several types of inverters. Depending on the application, you can choose between single-phase (230V) and three-phase inverters (400V) or hybrid inverters (useful for DC-side electricity storage). From a certain size of the PV system (e.g., 5kW), it makes sense to opt for a three-phase inverter.

In addition, it is necessary to distinguish between "String"-inverters and "Multistring"-inverters. You can build your PV system in such a way that you provide a separate inverter for each "String" (several "String"-inverters) or serve several PV-"Strings" with a "Multistring"-inverter.

4.2.4 Solar power meter (4)

Then an optional solar electricity meter or yield meter can follow, which measures the total electricity production of the PV system. This should not be confused with the normal house electricity meter, which only follows in point (6). However, modern inverters almost all have such a meter integrated, so you can usually save an additional solar electricity meter at this point.

4.2.5 AC overvoltage protection (5)

Surge protection must also be installed on the AC side of the circuit.

4.2.6 Main distribution (electricity box) with electricity meter (6)

Depending on the year of construction of your house or the electrical system of your house, you may have an analog or digital electricity meter installed. However, these electricity meters only measure the purchase of electricity, i.e., how many kWh of electricity are consumed from the public power grid. Depending on the amount of electricity and the price, you will then receive a bill.

 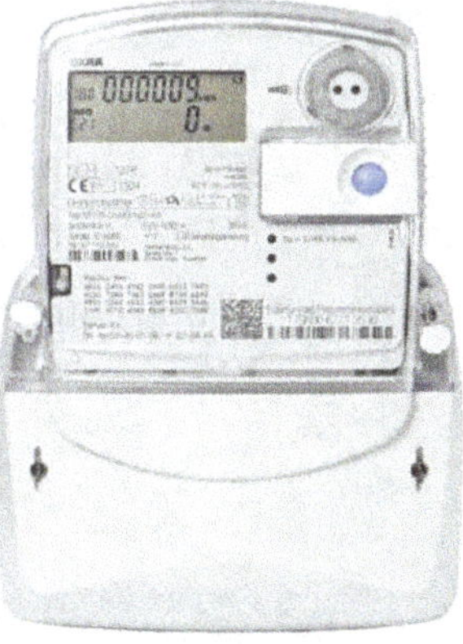

If you have decided in favor of a grid-connected PV system, however, you will now also need – in addition to such a supply meter – a feed-in meter. This feed-in meter measures the electricity that is fed into the public grid and is also used for subsequent billing. Normally, however, the old supply meter is simply replaced by a combined – bidirectional - meter. This takes over both counting directions. Visually, such a meter can hardly be distinguished from a digital supply meter. Only the designation bidirectional meter can provide the layman with the necessary information.

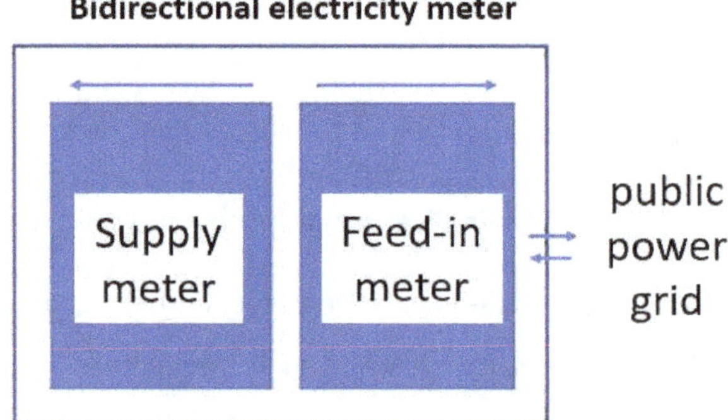

4.2.7 House connection to the public power grid (7)

This point connects a house to the public power grid. The house connection is used for both power purchase and feed-in.

4.2.8 Equipotential bonding (grounding) (8)

All components of the PV system and all loads must be grounded for safety. For this purpose, there is a local potential equalization (grounding bar) near the house connection. Fault currents are discharged into the ground via this grounding rail.

4.2.9 Consumer (9)

Consumers such as the washing machine or PC are connected normally via the house sockets.

4.2.10 Power storage (optional) (10)

The supposedly biggest problem in the context of generating one's own electricity is that electricity cannot always be consumed immediately when it is generated. Household appliances, e.g., refrigerators, require a continuous supply of electricity. Other appliances are often only switched on in the evening (e.g., TV) when the sun is no longer shining. If the electricity is fed into the public grid, this circumstance does not matter to the PV owner, since the electricity can be fed in at any time. However, if you want to achieve the highest possible self-consumption, you can solve this problem with an electricity storage system (battery storage). A high self-consumption makes sense, since one must buy electricity relatively expensively, for the self-produced fed-in electricity however only a comparatively small remuneration receives.

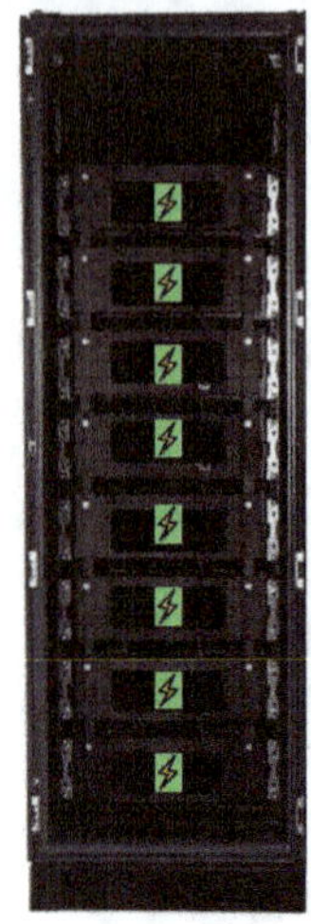  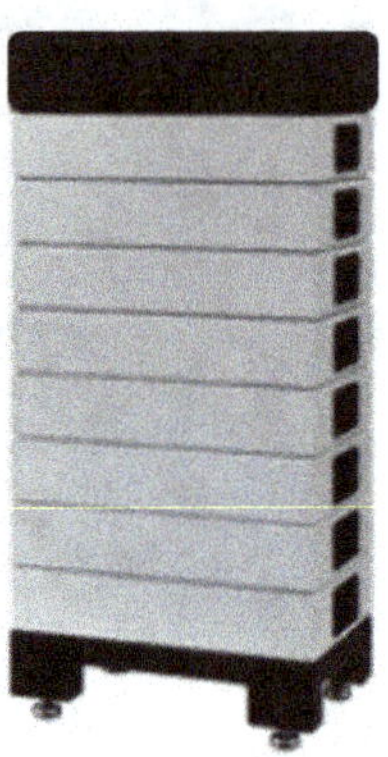

In order to be able to store the electricity produced in the battery storage, a charge controller and a battery inverter or voltage converter are also required. We will get to know these two components in more detail in **11.** and **12.** **<u>Note:</u>** In some electricity storage systems, these two components are already integrated and do not have to be purchased additionally.

4.2.11 Charge controller (only for current storage) (11)

The main task of a charge controller is to protect the battery storage during charging and discharging. The charge controller protects the battery from both excessive charging (overcharging) and excessive discharging (deep discharging) by regulating the charge current. Depending on the model, there is also a display for state of charge, charge current, battery voltage and temperature. There are generally three different charge controller types. 1. series controller, 2. "Shunt" controller (PWM), 3. MPPT controller ("Maximum Power Point Tracking").

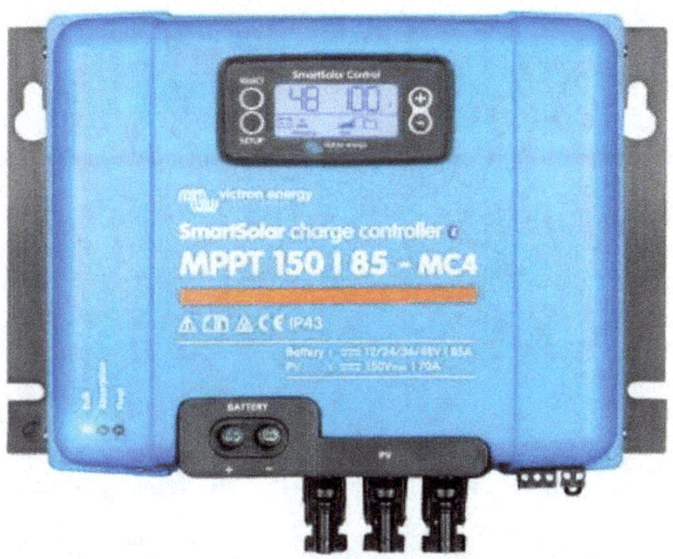

4.2.12 Inverter/rectifier or voltage converter for battery storage (AC-DC converter or DC-DC converter) (12)

Depending on whether you want to set up the photovoltaic system with an AC-coupled electricity storage system or a DC-coupled electricity storage system, you still need an AC/DC battery inverter (for AC-coupled battery storage) or a DC-DC voltage converter (for DC-coupled battery storage). Let's deal with both variants in the following. As already mentioned, there are also electricity storage systems that already have these components integrated.

AC/DC battery inverter in case of AC coupled battery storage:

This inverter is necessary because our battery storage is directly connected to the AC circuit in an AC-coupled system. However, the battery storage cannot be charged with alternating current, but in turn requires direct current. So, the inverter transforms the AC back to DC. This transformation from DC to AC (**PV inverter!**) and back to DC (**battery inverter!**) results in greater conversion losses than with a DC-coupled battery storage system.

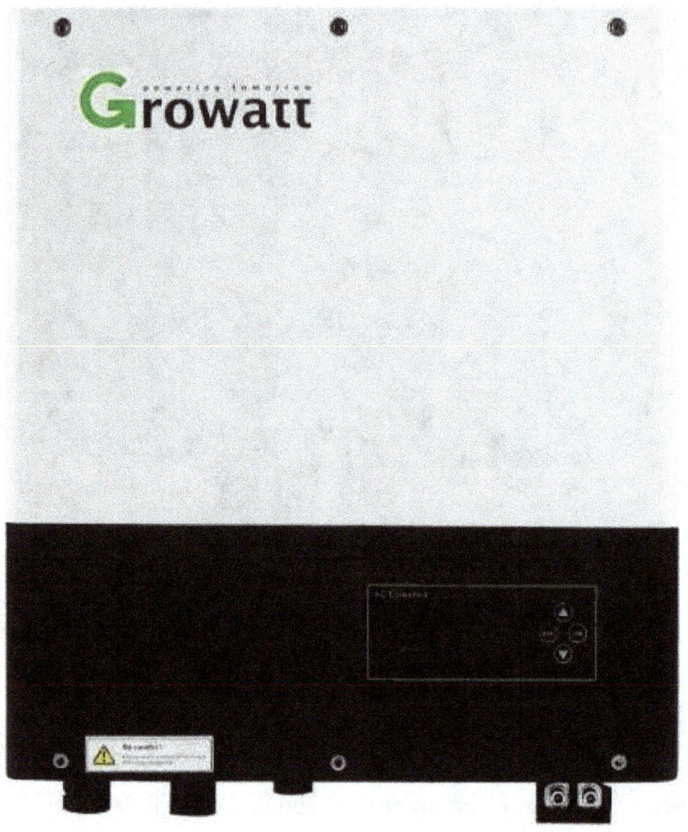

DC/DC voltage converter in case of DC coupled battery storage:

In DC-coupled battery storage systems, on the other hand, we already have the DC power we need (directly from the PV system). However, depending on the size of the PV system, circuitry and charge controller, we would still need a DC-DC voltage converter here, also called a DC-DC voltage converter, which reduces or increases the level of voltage (coming from the PV modules). Depending on the circuitry of the PV modules, the voltage may be too high for the charge controller or battery storage, so it must first be reduced with the DC-DC converter. In some cases this converter is also optional, we will come to this in the practical example.

4.3 Assembly of the plant

A PV system is usually installed on the roof of a house. This location is suitable due to the height (usually no shading by e.g., trees) and the orientation (one of the roof surfaces is often oriented to the south). If mounting on the roof surface is not an option, alternatives such as the roof of a garage, a hall, a carport, or even an open space can be considered. If you are planning the PV system for a new house, you can also integrate the PV system directly into the roof (so-called in-roof system) or even install solar roof tiles.

Solar roof tiles are a combination of roof tiles with an integrated photovoltaic module. From a greater distance, these can hardly be distinguished from normal roof tiles. There are various suppliers and designs for this. The tiles are hooked into the roof battens like conventional tiles and are also wired together immediately. Solar roof tiles are still relatively expensive.

In the following, we will now look at the mounting of conventional PV modules on a house roof. Depending on the type of roof, there are different mounting systems. **Note:** The roof must be able to bear the load of the PV modules (in addition to any snow load), i.e., it must be structurally designed for this. If you are unsure, you can consult a structural engineer. The drainage of rainwater must also not be obstructed by the PV system, otherwise consequential damage to the roof may occur.

Generally, the PV modules are mounted on a suitable substructure of aluminum profiles, which in turn are anchored in the roof. Depending on the roof shape (gable roof, flat roof, monopitch roof, hip roof ...) different substructures are used and depending on the roof tile (tile, slate ...) different mounting methods. Let's look at two examples.

4.3.1 Saddle roof mounting

The PV modules are mounted parallel to the roof surface in the case of a gable roof or generally in the case of sufficiently steeply sloping roof shapes.

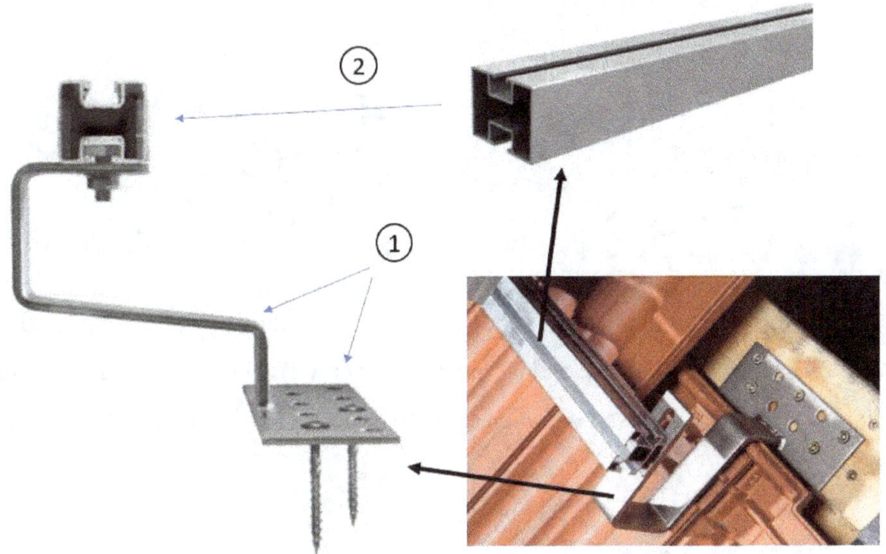

Below the roof tiles, several so-called roof hooks or roof anchors **(1) are** screwed onto the roof battens. Aluminum profiles **(2)** are then mounted between these roof hooks to support the PV modules. See picture above.

Depending on the type of roofing (tiles, shingles, plain tiles ...) you need the designated version of the roof hook. Hanger bolts **(1) are** used for trapezoidal corrugated roofs, version **(2)** for slate roofs, version **(3)** for plain tiles. It is also possible to replace the tiles in the area of the roof hooks with sheet metal tiles **(4)** and **(5).** This can prevent the tile from breaking underneath the roof hook. Conventional tiles can break due to the load acting on the tile below the roof hook – over time.

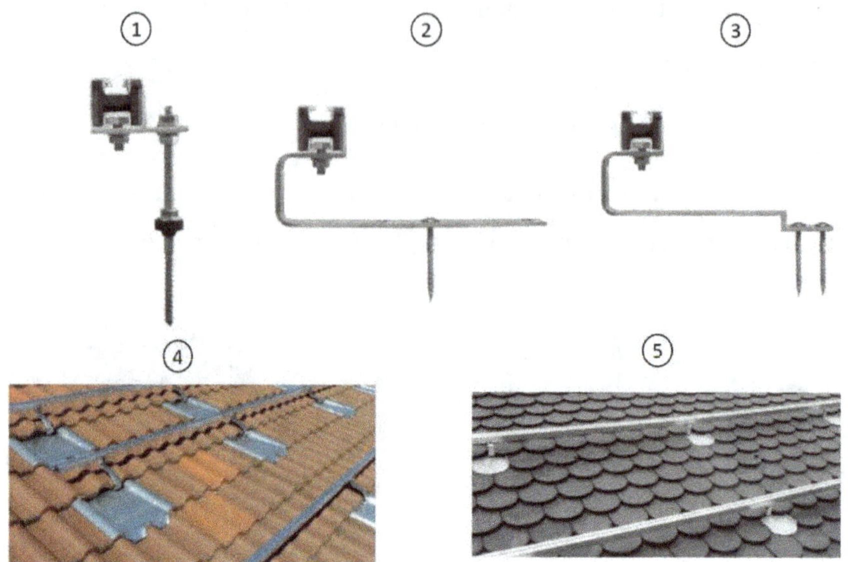

The PV modules are then clamped onto the aluminum profiles using module clamps (mounting material: sliding blocks and screws). There are module center clamps **(1), which are** used between two PV modules, and module end clamps **(2)** for the edge areas.

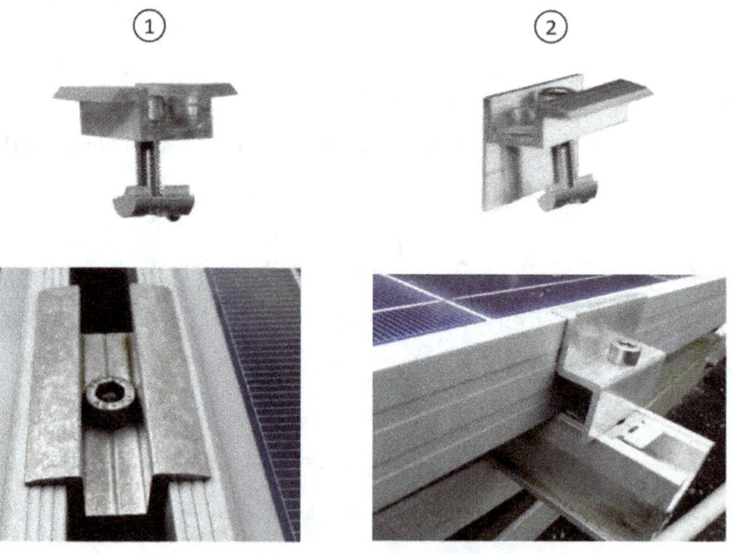

4.3.2 Flat roof mounting

On a flat roof or a roof without sufficient slope, the mounting situation is somewhat different. Due to the lack of slope, the PV modules are elevated for better exploitation of solar radiation. Elevated means that a frame construction in the form of a triangle is constructed below the PV modules. This creates a sufficient slope. In addition to better solar utilization, a self-cleaning effect takes place due to an angle of 25° or more, as dirt and debris can be washed away better when it rains.

Elevation also takes place with ground-mounted systems. Here you will also find a more complex substructure.

In addition to or instead of the roof, the house facade can also be used for a PV system. Thin-film modules are particularly suitable here.

4.4 Acceptance and commissioning

If you want to install a PV system on your roof, the guidance of a PV expert or electrician is of great importance for your individual project. In this book, you will learn the basics about PV systems and also learn from practical examples how to plan a PV system, but what your individual project will ultimately look like should be reviewed again by an expert. Be aware that improper planning, design or installation can cause significant damage to property and people. The most common damage to roof, people, or the entire house is caused by fire, storm, snow load, and lightning. Therefore, be sure to get additional advice for the planning and installation of your individual PV system! If you decide on a PV system with grid feeding, you will have to hire an electrician anyway (at least to connect the PV system), depending on the country. The installation of the modules, on the other hand, can be done by yourself. Find out about the laws and reporting requirements that apply in your country. In addition, you must have your electricity meter replaced when feeding PV electricity into the grid if you have not installed a bidirectional meter.

4.5 Special form: Mini PV systems or balcony power plants

Apart from permanently installed PV systems with several modules, there has also been the possibility of using plug-in small PV systems for some time. These PV systems consisting of only one or two to three modules **(1)** with integrated module inverter or additional micro inverter **(2)** are often also referred to as balcony power plants or guerrilla PV systems. The name balcony power plant already suggests that one can install this type of PV system mainly on a balcony or a terrace. Thus, even as a tenant, you can generate your own electricity. With the help of a special feeding socket **(3)** (exchange with a normal socket by an electrician) the plant can be connected directly to the electric circuit of the apartment, thus reducing the power consumption. This special socket is more robust in design than a normal socket and serves to protect against fire or short circuits.

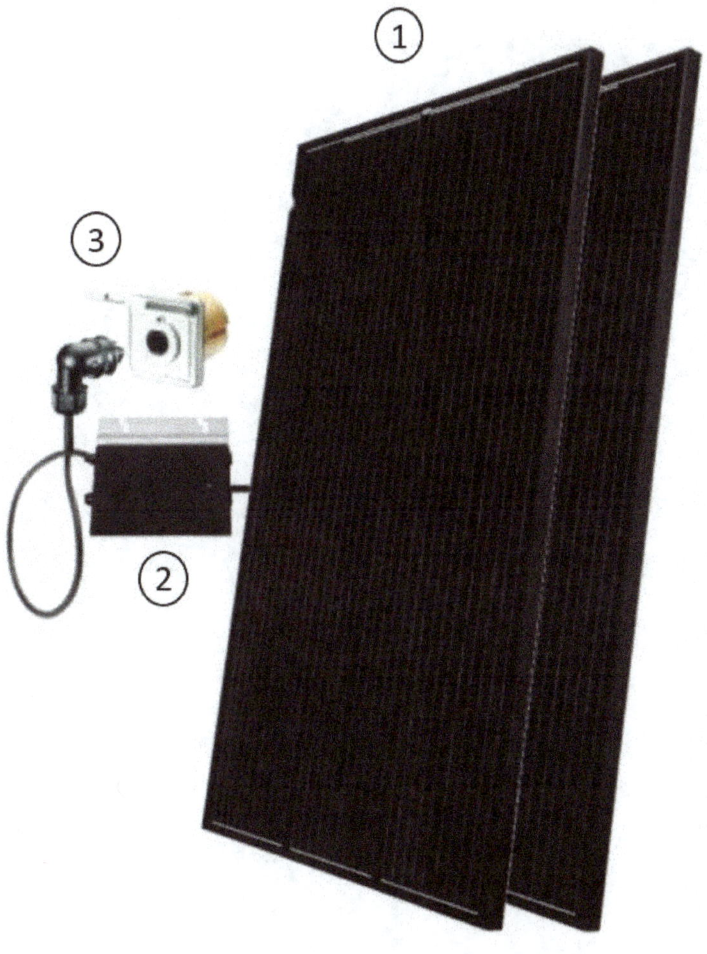

If the mini PV system does not provide more than 600 watts of output power and is certified according to safety standard DGS 0001, the connection can also be made to a standard household socket, according to the German Solar Energy

Society (DGS). In general, depending on the power consumption and size of the mini PV system, you must also in any case inquire whether the built-in electricity meter for the apartment is modern enough so that it does not run backwards if more electricity is fed in than is consumed. In any case, the installation of the mini PV system on the balcony may have to be approved by the landlord or property management. It is best to ask your landlord about this.

5 Practical example: Island system for mobile home or Tiny House

5.1 Planning, component selection and connection of the stand-alone PV system

In this chapter, we will look in detail at a practical example of how the planning, component selection and connection of an off-grid system for e.g., a mobile home or even a Tiny House could look. We will do this step by step.

In this example, we assume that the installation site (e.g., roof surface) can bear the load of the PV system and that the orientation cannot be changed or, in the case of the mobile home, that it faces south. Furthermore, we assume a pure summer operation (with daily power consumption) of the PV system.

5.1.1 Step 1: Estimating the power consumption

This step is about calculating the power consumption of the Tiny House or RV. We have to do this to get the correct number of PV modules for further planning. After all, the PV system should not be oversized, but also not undersized. For example, we assume that we want to run a refrigerator, a small TV and lighting in our RV or Tiny House. We would also like to have an additional 230V outlet for charging a notebook or cell phone. Of course, you can also add individual devices. We note down these consumers in a table, with a column for the name of the device, with a column for the energy demand [W] and with a column for the usage time [h]. After you have listed all the appliances, note the number of hours these appliances are on each day. The refrigeration compressor in the refrigerator does not run continuously 24h, but only switches on when the actual temperature exceeds the desired set temperature. Here you can take a look at the data sheet to get the averaged power consumption. We plan on about 4 hours of operation at 50W per day. We schedule the TV for 2 hours of daily use, the wall outlet for 3 hours, and the lights for 4 hours (summer months). In general, it is better to plan a little more so that there are enough reserves. In the third column, we still note the wattage of each appliance based on the information on the appliance's nameplate. We can calculate the total energy requirement in watt-hours by multiplying the wattage of the devices by the hours required to operate them and then adding up the individual results. This could then look like this, for example:

Device	Energy demand [W]	Operating hours [h]	Total energy demand per day [Wh]
Refrigerator	50	4	200
Lighting	20	4	80
TV	40	2	80
Power socket (PC)	90	3	270
			= 630 Wh = 0.63 kWh

After we have calculated the total load in watts and the total energy demand in kilowatt-hours, we can estimate the capacity of our solar system and battery storage. But before that, we have to take care of an essential point in the planning – the system voltage.

5.1.2 Step 2: Plan the system voltage

Before we get to the selection of our battery storage, we deal with the system voltage for our stand-alone PV system. This step is crucial because the chosen system voltage influences the selection of the following components. Basically, a stand-alone PV system can typically be operated with 12V, 24V or even 48V system voltage. System voltage refers to the voltage that flows in the system (i.e., between the interconnected PV modules via the charge controller into the battery storage).

The larger the PV system, the higher the system voltage should be in order to save costs (especially for the charge controller). From 500 Wp total power of the planned PV system, you should consider switching from 12V to 24V. A higher voltage also has the advantage that longer distances (cables) are associated with fewer losses. In addition, one needs less thick cables (smaller cable cross-section) at higher voltages, since the current intensity is reduced here for the same power ($P = U \times I$). If you stay below 500 Wp total power of the PV modules and maybe even have some 12V consumers, you can plan a 12V system. How the different system voltages affect in detail, we will see during the next steps. In our case, we decide for a 24V system.

5.1.3 Step 3: Calculate the battery storage size

Since we are planning an off-grid PV system, we require a battery storage to sufficiently cover our electricity needs even if there is no sun or the sun is too weak (due to: clouds, night ...). To do this, we need to calculate the size of the battery storage. For this, we use the determined total energy demand from step 1. In addition, we plan a reserve so that our power supply also works if the sun does not shine for several days. We plan this reserve with 2 days. This means that the power supply can be maintained for 2 days even if there is very little sunshine (i.e., the battery is hardly charged). For this, we multiply our calculated total energy demand of 630 Wh by the factor 2 (2 days self-sufficiency): 630 Wh x 2 = 1260 Wh. In this step, one can also include a capacity reserve of e.g., 30-50 % to compensate for line losses and a lower sunshine duration. In this case, the value calculated so far would be multiplied by 1.3 (30% reserve) or 1.5 (50% reserve). For our pure summer operation with high sunshine duration, we do without this in this example. However, for your individual project it is recommended to consider a reserve of at least 15%.

Now, to get the capacity of the battery storage, we have to divide our result in Wh by the battery voltage (we want to use a system voltage of 24 V). Then we get the required battery capacity in Ah (amp-hours). That is: 1260 Wh / 24 V = 52.5 Ah.

In addition, we must consider that the battery capacity must not fall below a certain value in order not to discharge the battery storage too much and thus possibly damage it. A value between 50 and 80% of the battery capacity is acceptable for this. Lead-acid batteries, for example, may only be discharged to 50% (or according to manufacturer specifications for special AGM types). In this case, we must therefore multiply our previously calculated battery capacity in Ah by a factor of 2: 2 x 52.5 Ah = 105 Ah.

If we choose a lead acid battery, we should definitely choose a special solar battery or a lead acid battery with AGM or gel technology. The difference between these two types of batteries is the bonding of the electrolyte. If we choose a more modern and compact technology, e.g., a lithium battery storage, we only have to consider 10-25% residual capacity instead of 50% for battery discharge. However, lithium battery storage is significantly more expensive than conventional lead-acid batteries. The advantages and disadvantages are clearly summarized in the following table:

	GEL	AGM	Lithium
Charging cycles	500 - 1800	400 - 1500	2000 - 5000
Lifetime	10 - 12 years	7 - 12 years	Greater than 12 years
Advantages	Higher depth of discharge, moderate costs	High current, low cost, low self-discharge	High current, very deep discharge possible, high efficiency,

			compact, very low self-discharge.
Disadvantages	High space requirement	Lower depth of discharge,	High cost

In our case, we opt for an AGM battery because we shy away from the higher costs of a lithium battery investment. Although lithium batteries are worthwhile in the long run, they are three to five times more expensive than AGM batteries in the beginning.

So, we have now calculated a value of 105 Ah for our battery storage. Now we could use a single AGM battery with 105 Ah capacity for this. This would then have to have 24 V, but most AGM batteries only have 12 V. To achieve the required 24 V, we have to connect two batteries in series (series connection: voltage adds up, current remains the same). If the capacity of the selected batteries is not sufficient, we can also connect additional batteries in parallel (parallel connection: current strength adds up, voltage remains the same). In our case, for example, we choose two "12V 110Ah Deep Cycle AGM" batteries from "Victron Energy", which we connect in series. We then get 24V 110Ah. 105Ah are needed, which means that 5Ah remain as reserve. By the way, we will see in detail how we connect all components in the last step.

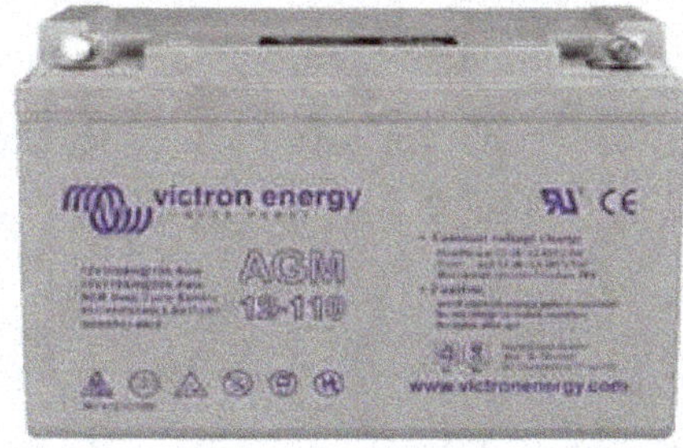

https://greenakku.de/Batterien/AGM-Batterien/Victron-Energy-12V-110Ah-Deep-Cycle-AGM-Batterie::1179.html

<u>Note:</u> With a system voltage of 12V, we would have obtained a value of 210 Ah for the battery capacity with 1260 Wh / 12 V = 105 Ah and with 50% residual capacity (2 x 105 Ah). For this, we could have planned, for example, a larger battery from "Victron Energy" with 12V 220 Ah. If you compare the prices, you will pay about (!) the same price in both cases. However, for the solar charge controller, the price difference will be significantly different.

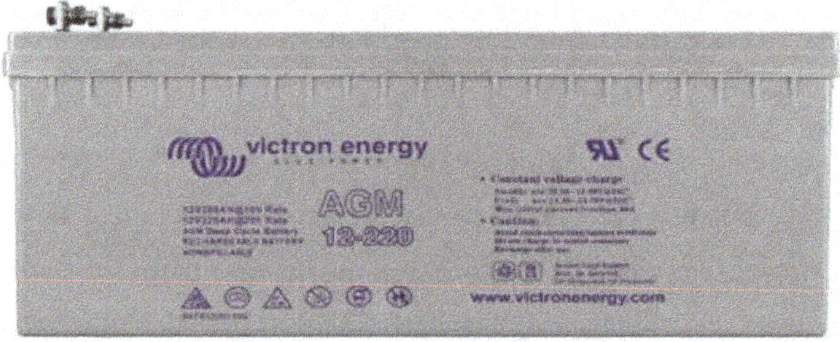

https://greenakku.de/Batterien/AGM-Batterien/Victron-Energy-12V-220Ah-Deep-Cycle-AGM-Batterie::1189.html

If you want to choose the lithium technology, a lithium alternative would be, for example, the lithium battery from "Liontron" with 25.6V and 100A. By the way, you can easily find 24V and 48V batteries in lithium battery storage, so a single battery is sufficient here. In this case, the capacity [Ah] of the lithium battery storage would even be dimensioned a bit too large because we do not have to consider 50% residual capacity with lithium, but e.g., only 20% (see previous calculation and explanations). With a factor of 1.25 instead of a factor of 2 we would get 65.6 Ah battery capacity (1.25 x 52.5 Ah = 65.6 Ah), i.e. a 65-70 Ah lithium battery storage would also be sufficient for this. A lithium battery will still be a lot more expensive, but the price-performance ratio is much better here.

5.1.4 Step 4: Determine size and number of PV modules

By now, we know what our total energy demand and battery storage is. To ensure that our solar batteries are also charged, we take care of the selection and dimensioning of the PV modules in this step. Before we can do that, we need data for the installation site, the orientation of our system and the optimal installation angle. So that it does not become too complex, we assume that our Tiny House or mobile home is permanently located in one place (e.g., camping site) and is <u>not</u> driven to different places.

As we already know if we are in the Northern Hemisphere (Europe, USA ...), we should orient our PV modules as closely as possible to the south. In addition, there should be no shading of the modules throughout the day. The best way to try this out at your chosen location is to simply observe the course of the shadows over several days.

In order to obtain the optimal installation angle, we have to take into account the location (latitude) and the season (in our case: summer mode). To do this, we can either take a look at a map or read out the exact data with the help of two online tools that we will need in the further course. This is for example the tool "PVGIS", with which one can determine parameters for PV plants for a certain location. This tool can be found on the following website:

https://re.jrc.ec.europa.eu/pvg_tools/en/

Alternatively, one can use the website: https://globalsolaratlas.info/ to find out solar parameters for a specific area. However, we will handle the use of this internet site in the next chapter.

We first deal with the tool "PVGIS". We can enter the desired address of the location to be viewed in the area at the bottom left. In our example, we use the location Munich in Bavaria. With a click on "Go" the location is selected.

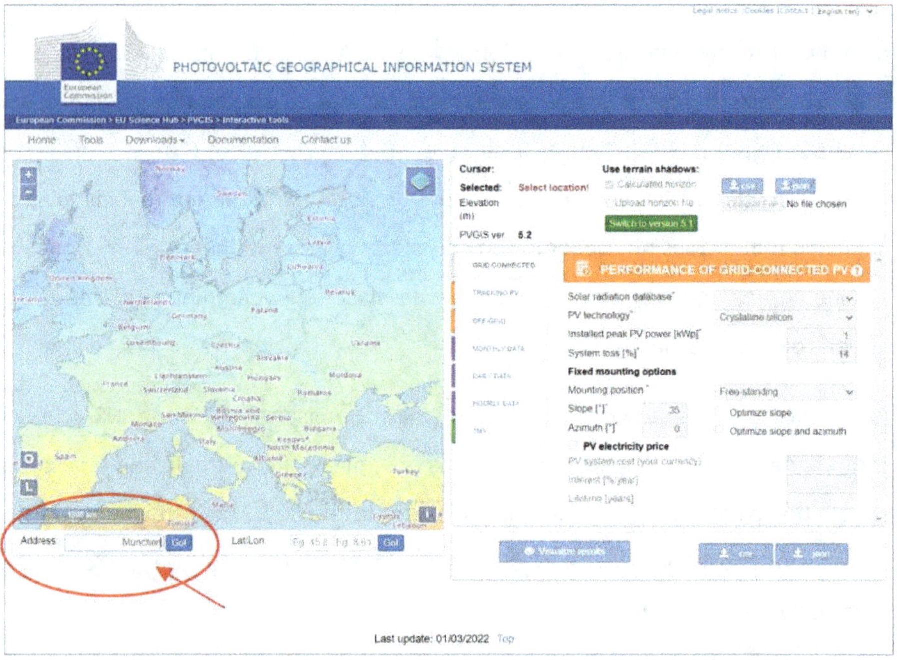

We then get a detailed view of the location and the selected latitude and longitude, as well as an indication of the location altitude in the marked area.

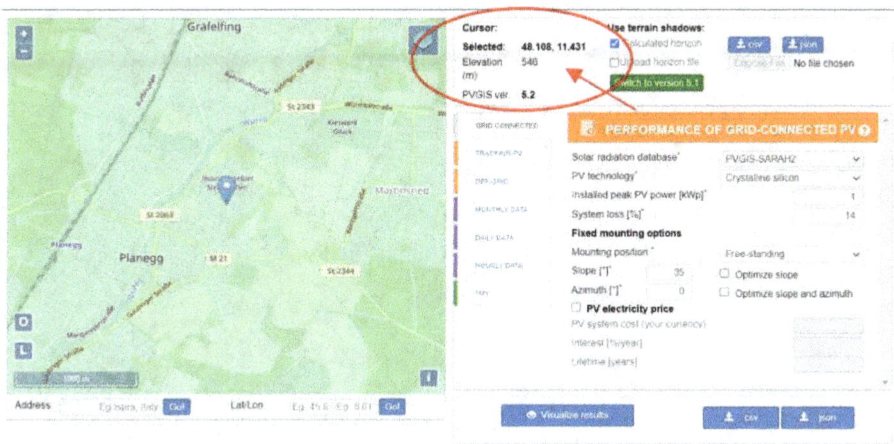

Thus, in the first step, we obtain the required latitude, namely approx. 48° for the Munich location. For the calculation of the optimal installation angle of our PV modules, we then use the following formula: 90° minus peak of the sun at "x" degrees. We obtain the peak of the sun with 90° - (48° - 23°) = 65°. This means for the **optimal installation angle**: 90° - 65° = **25°**. The 23° correspond to the approximate angle of inclination of the earth. We have thus calculated that with an installation angle of 25° and orientation to the south, we obtain the maximum energy yield on the day of the summer solstice.

Note: If we would prefer to operate the island system all year round, we would have to design the system for winter operation. In this case, we obtain the optimum installation angle if we calculate the peak of the sun as 90° - (48° + 23°) = 19°. The optimal installation angle in this case would then be: 90° - 19° = 71°, i.e., significantly steeper.

Now that we have calculated the optimal installation angle, we can calculate the yield of our PV system in the tool "PVGIS" by entering some required parameters. To do this, we change the type of the system in the middle area to "OFF-GRID", since we are planning an off-grid stand-alone system.

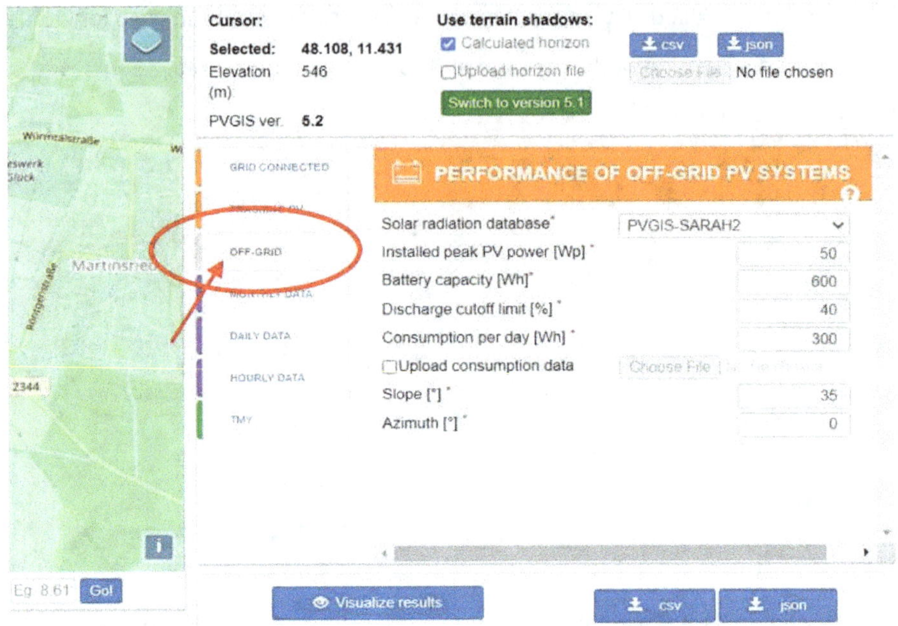

Now let's start entering our values. We start with the angle of incidence ("Slope") and the orientation ("Azimuth") in the lower part of the entry fields. At "Slope" we enter the calculated setup angle of 25° and at "Azimuth" we enter 0° (corresponds to orientation to the south). So far, we still have the total energy demand of 630 Wh calculated in the first step, which we enter at "Consumption per day [Wh]". As "Solar radiation database" we leave the database "PVGIS-SARAH2", which contains data from 2005 to 2020. We have also already calculated the battery capacity (actually 105 Ah, but we use a 110 Ah battery storage). We have to convert this Ah into Wh (110 Ah x 24V = 2640 Wh) and enter it at "Battery capacity [Wh]". At "Discharge cutoff limit [%]" we enter the 50% depth of discharge measured in the previous step.

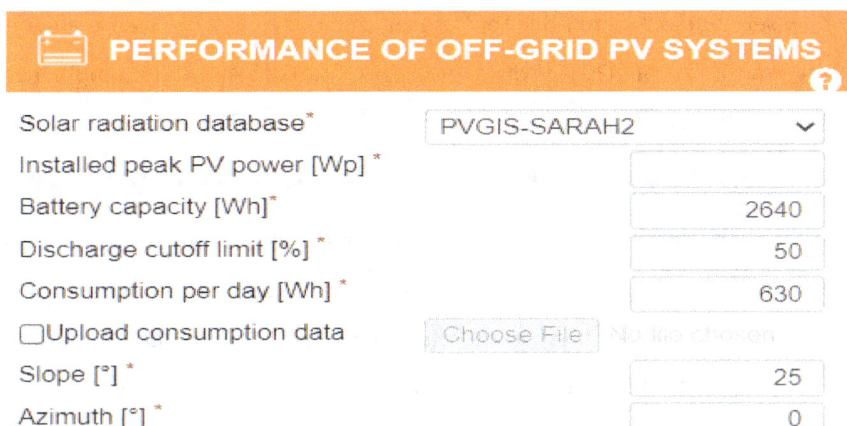

One parameter remains, namely "Installed peak PV power [Wp]". We want to find out this value, to do this we first simply enter any value, e.g., 500 [Wp] and display the results. We can then see if we need a higher total PV module power, or even if it is already too high. To do this, we click on "Visualize results" and then on the button "Performance" to be able to assess the days with a full or empty battery.

We now see that in the months April to September we would only find an empty battery a maximum of 1.25% of the days in the month of September (empty means here due to the discharge limit entered: 50% discharged). This means that with PV modules with 500 Wp total power, we would already be pretty well-equipped.

Based on the month of September, this would mean that we would probably only have to manage with an empty (= 50% discharged) battery for a fraction of a day (30 days x 1.25% = 0.375 days). In the winter months, as can be seen from the red bars, it looks quite different. However, as already mentioned, we only consider summer operation (April-September). In order not to have to manage a single day with an empty battery, we simply increase our total power in 50 Wp steps until we no longer see a red bar between April and September ("Days with battery empty" below 0.5 %). In this example, this would already be the case at 550 Wp.

So, we need a total PV power of about 550 Wp. What does Wp mean again? Wp stands for "Watt peak" and indicates the maximum power of a solar module when it is operated under optimal conditions (cell temperature = 25 °C, irradiance = 1000 W/m², air mass = 1.5). This value can be used to compare PV modules with each other.

PV modules are available on average between 50 Wp and 400 Wp, depending on size and design.

In our case, for example, we could install either two 360 Wp **monocrystalline** "BlueSolar" solar modules from Victron Energy (available here, for example: https://greenakku.de/Solarmodule/Solarmodule-ab-200Wp/Victron-BlueSolar-Solarmodul-Monokristallin-360Wp::3678.html), or alternatively two 330 Wp "BlueSolar" modules in **polycrystalline** design (available here, for example: https://greenakku.de/Solarmodule/Solarmodule-ab-200Wp/Victron-BlueSolar-Solarmodul-Polykristallin-330Wp::3622.html).

This gives us even a little more power than desired (can be seen as a reserve). We would then get either 2 x 360 Wp = 720 Wp or 2 x 330 Wp = 660 Wp instead of the required 550 Wp. If you don't want such a high reserve, you can also choose smaller modules with e.g., 50 Wp each and connect 11 of them together to get the 550 Wp.

It is worth noting that there are also ready-made packages with PV modules, battery storage, charge controller ..., which can partially create a price advantage. When searching, orient yourself on the required parameters (Wp for PV modules and kWh or Ah for battery storage). If a charge controller is also included in the package, read the next step first, or in general, it is best to read through all steps completely before you decide.

5.1.5 Step 5: Charge controller

Since we now know the capacity of the battery storage and the number of PV modules for our example project, we can continue with the charge controller. What is the purpose of a charge controller? This component ensures that the battery storage is charged with the correct voltage, it also prevents overcharging of the battery storage, in addition, it protects the battery storage from deep discharge and ensures a long service life of the system.

Basically, there are two different types of charge controllers, one is the "PWM" type and the other is the "MPPT" type. Let's take a brief look at the differences.

PWM

MPPT

The abbreviation "PWM" means pulse width modulation. Until the battery is fully charged, this charge controller virtually lets as much current flow as necessary to fully charge the battery. Only when the battery is fully charged does the charge controller continuously switch the current to the battery on and off (PWM technology) so that the battery maintains a constant voltage. So, you can think of it simply as a kind of monitor that – depending on the voltage state of the battery – turns a switch on or off. A charge controller with pulse width modulation should be used whenever the voltage from the interconnected PV modules is similar to the voltage of the battery storage (e.g., PV circuit supplies ~ 12V & battery storage is planned as a 12V system). A PWM charge controller is simpler and cheaper than an MPPT charge controller and is therefore mostly used in smaller PV systems.

The abbreviation "MPPT" stands for "Maximum Power Point Tracking". The "Maximum Power Point" is the point at which the ratio of current to voltage is optimal for maximum power generation. This point changes depending on the solar radiation, the "MPPT" charge controller can take this into account. This charge controller is quite a bit more expensive than a PWM charge controller, but it is also much more efficient. Especially for larger PV systems and a higher PV voltage, you should choose a "MPPT" charge controller.

You may also be missing a DC-DC converter in our previous setup, as described in one of the previous chapters. However, as mentioned, this is only needed if the voltage between the PV module circuitry and the battery storage does not match. If that were the case, we would need a DC-DC converter if we decided to use a PWM charge controller. On the other hand, if we select a "MPPT" charge controller, most of the devices already have a DC-DC voltage converter built in internally, so we can save an additional component.

So, how do we calculate the size of the charge controller for our system? This is not so complicated.

Firstly, we need to select the charge controller in terms of the required current [A]. To do this, we simply need to divide the total power of our PV modules [Wp] by the voltage [V] of the battery storage. In our example, selecting the 2 x 330 Wp PV modules, this would mean: 660 Wp / 24 V = 27.5 A. The solar charge controller should therefore deliver at least up to 28 A charge current. This calculation also shows why it makes more sense to choose a higher battery voltage for larger PV systems or higher power [Wp]. If we had picked a 12V system voltage at 660 Wp, we would need a much higher charge controller at 660 Wp / 12 V = 55 A (about twice as expensive). For a 1500 Wp system, for example, we would even need a 125 A charge controller with a 12V battery voltage. Such large charge controllers are very expensive and can be avoided by planning the system voltage higher. At 24V battery voltage, for example, a 62.5 A charge controller would be sufficient and at 48V battery voltage already a 31.25 A charge controller. These are significantly cheaper.

On the other hand, we must check whether the selected charge controller is designed for the maximum total current of our PV module connection. To do this, we look for this value in the data sheet of the charge controller and compare it with the sum of all PV module current strengths connected in parallel. Because, as we know, the individual currents add up in a parallel connection, the voltage remains the same. Of course, the voltage of the module circuit must also match the input voltage of the charge controller. This is primarily important for a series connection of PV modules because here – as we also already know – the individual voltages add up, but the current strength remains the same.

As MPPT charge controller would be in our case, for example, the "BlueSolar MPPT 100/30" solar charge controller with 12 V or 24 V battery voltage (adjusts automatically) and 30A charge current come into question. Up to 100V PV voltage can be applied here.

https://greenakku.de/Ladegeraete/Solarladeregler/MPPT-Solarladeregler/BlueSolar-MPPT-100-50-Solarladeregler-12-24V-50A::615.html

5.1.6 Step 6: Inverter

To be able to supply 230V consumers with our PV system, we still need an inverter. Here, we should choose a pure sine wave inverter, as this best represents the household current. There are also rectangular inverters and trapezoidal inverters. To size the inverter, we need to add up all the loads to determine the maximum power. Of course, not all loads will always be on at the same time, but we assume the worst-case scenario (all loads on at the same time) for sizing. In our case, this would look like this:

P_{max} = 50 W + 20 W + 40 W + 90 W = 180 W

For this, we remember the planned consumers with the following values:

Refrigerator	50 W
Lighting	20 W
TV	40 W
Power socket (PC)	90 W

Here we only have to consider the consumers that are to be connected to the 230V power supply, i.e., to the inverter. We can omit devices with 12V or 24V voltage supply here because they cannot be supplied via the inverter, but directly (or via voltage converter and fuse protection) via the battery storage or from the solar charge controller.

Since we also have to take losses into account when converting direct current to alternating current, we plan a reserve – depending on the inverter. For example, we assume that our inverter has an efficiency of 88% (see the manufacturer's data sheet). This generally means: P_{max} / efficiency = required power. In our case, this means: 180W / 88 % = 204.54 Watt. You should not dimension the inverter too tightly and rather choose a larger one. Otherwise you will have to buy it again when expanding the PV system. The temperature also plays a role in the power (see data sheet).

In our case, for example, a "Phoenix" 24/500 inverter would come into question. The number 24 stands for 24V, the number 500 for VA (1 VA = 1 Watt), thus for the power (we need at least 204,54 Watt). This inverter converts 24V into 230V pure sine wave voltage.

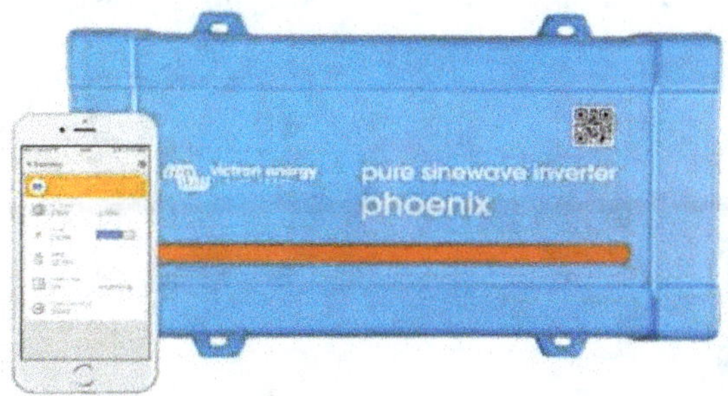

https://greenakku.de/Wechselrichter/Offgridwechselrichter/Victron-Phoenix-Wechselrichter/Phoenix-Wechselrichter-24-500-230V-VE-Direct::816.html

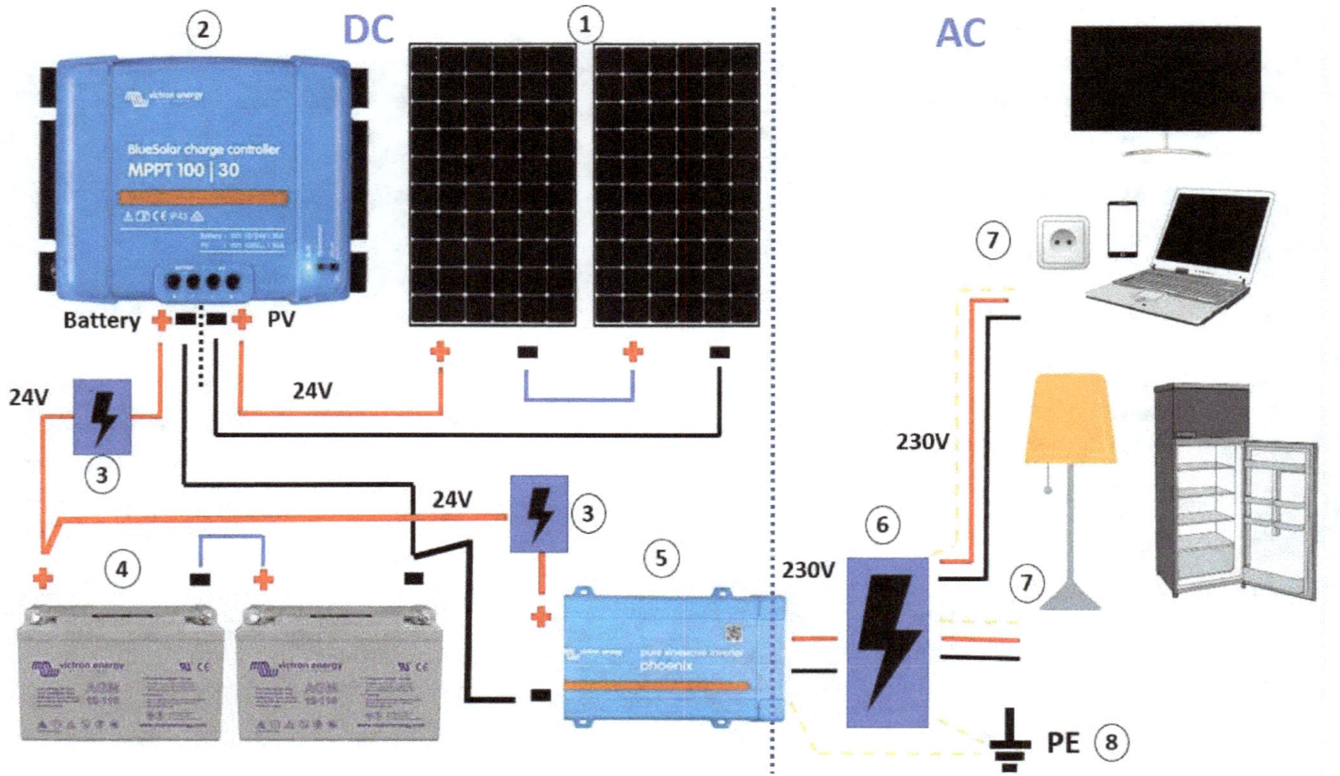

DC
AC
2
1
BlueSolar charge controller
MPPT 100 | 30
victron energy
Battery
PV
24V
24V
24V
3
3
4
7
230V
230V
6
5
phoenix
7
PE
8
5.1.7 Step 7: Wiring or setting up the system

Now we come to the wiring of our components for our mini PV system. As shown in the figure, the current comes from the PV modules, which we connect in series **(1)** via special solar cables to the charge controller **(2)** (connection: "PV = Photovoltaic"). The battery **(4)** is also connected in series (2 x 12 V = 24 V) and charged via the charge controller **(2)**. A DC fuse **(3) is** installed in between. The inverter **(5)** draws DC current - via a fuse **(3)** – directly from the battery storage. Both fuses **(3)** should be installed as close as possible to the pole ("+") of the battery. After conversion to AC, an AC fuse box with circuit breakers should be installed. Suitable grounding **(8)** should also be considered. Finally, the current then reaches the consumers **(7)**.

Make sure that the capacities of the batteries you connect in series are the same (here, for example, 110 Ah each). Also note that you select the wire cross-sections and fuses correctly. Be sure to get expert advice from an electrician if you are unsure here for your individual project. In the worst case, improper design could otherwise result in fire or personal injury!

The wire cross-section depends on some parameters such as length, amperage, and voltage and can be calculated using the following formulas:

Calculation of the line cross-section for **direct current:**

$$A = \frac{2 \cdot l \cdot I}{\gamma \cdot U_a}$$

Calculation of the line cross-section for **single-phase alternating current:**

$$A = \frac{2 \cdot l \cdot I \cdot \cos\varphi}{\gamma \cdot U_a}$$

where A = conductor cross-section [mm²], l = length of conductor [m], I = current [A], γ = conductivity conductor [S/m], U_a = permissible voltage drop of cable in %, cos φ = power factor.

Alternatively, you can search online for a suitable program for calculation, in which you can simply enter the parameters.

In a smaller 12V or 24V system or a PV system with a charge controller with a lower charge current, there is often an additional current output directly on the charge controller. Smaller consumers can be connected here. This should be distributed or fused via a DC distributor **(9)**. The setup could be modified as follows for this purpose. The AC side (alternating current) can – as before – be additionally considered if required.

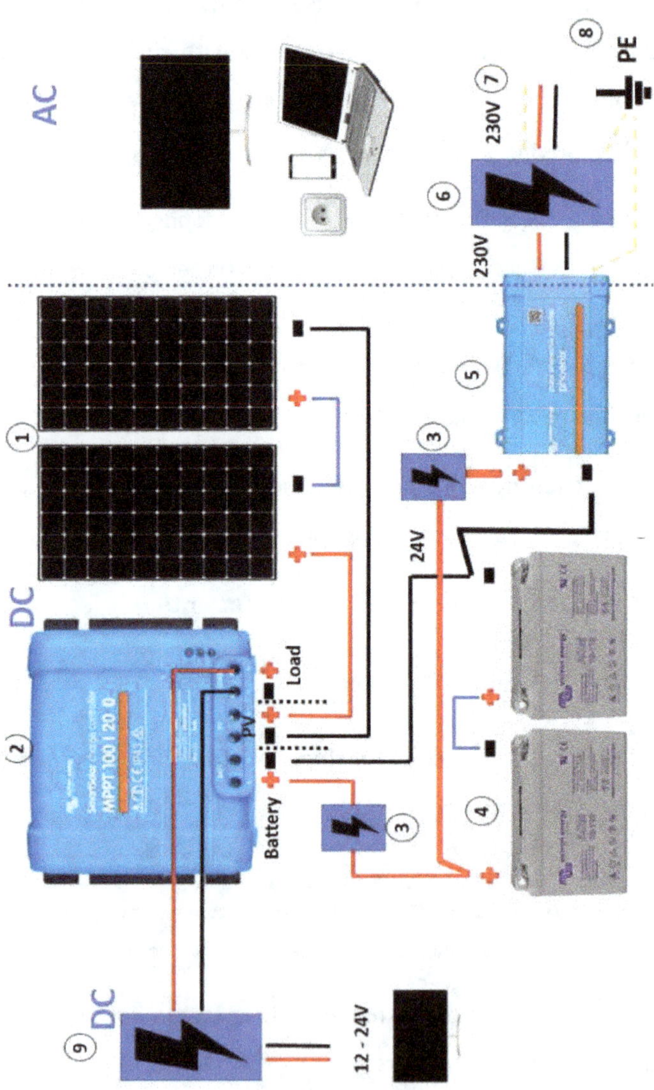

Devices with a high starting current (hairdryer, pump, refrigerator), which can be operated with 12V or 24V (mostly camping equipment), should however not be connected to the "Load"-output of the charge controller, but directly to the battery storage – via a DC fuse box. Connecting to the "Load"output of the charge controller, on the other hand, has the advantage that it disconnects when the battery charge is too low and thus protects against deep discharge.

6 Planning a photovoltaic system for your own home

We have learned a lot so far about planning and designing a stand-alone PV system. In this chapter, we will look at the planning of a grid-connected photovoltaic system for a house.

As already mentioned, the support of a PV expert or electrician is of great importance for your individual project. A PV system with grid feed-in must be connected by an electrician anyway. Therefore, please regard this chapter as a purely informational basis.

In the following, we will again proceed step by step. Some aspects will be similar to the planning of an island plant, other aspects will be new.

6.1 Step 1: Check installation site or roof surface

The first step in planning a PV system is to consider the area of the roof or the place where the solar system will be installed. It is not necessary to install a photovoltaic system on the roof of the house, you can also install your system on the roof of the garage or near the ground. In this case, however, one must keep in mind that obstacles in the surrounding area (trees, other houses, one's own house in the case of the garage roof, ...) are likely to contribute to the shading of a low-lying location. The location should therefore be as high as possible or protected from shadows and – assuming we are in the Northern Hemisphere - face south, as this is where the sunshine duration can be maximized. However, a deviation of up to 30 percent in the direction of the east or west can still be accepted. Depending on how you want to use the generated electricity (feed-in, own use, mixed use), you can consider how you want to proceed. One can either install PV modules on the entire usable area of the selected site to obtain the maximum possible power for the site, or alternatively, install only enough PV modules to meet the desired (e.g., one's own) electricity demand. For both cases, in this first step you should get an idea of the maximum available area (e.g., roof area) by measuring it. On the one hand, you can do this manually with a meter measure, but on the other hand, you can also use suitable planning software on a PC to carry out or calculate this. For this purpose, there is, for example, the possibility of having the roof area measured with drones. In the case of a flat roof, it is again relatively simple; here, you can assume the footprint of the house as the roof area. If the roof pitch is known, the available area can also be calculated from this. There are many online calculation programs for this purpose. We will look at an excellent calculation program in detail shortly. One must also take into account, for example, the chimney or other roof elements that reduce the available installation area. Of course, it is also necessary to check whether the roof can support the additional PV load. For this, one should consult an expert (e.g., structural engineer). With a modern roof, however, there should not be any problems here as a rule.

Before we continue with the further steps for planning our photovoltaic system, we will briefly look in more detail at determining the solar radiation for a specific location.

6.2 Step 2: Check irradiance

Basically, the generation of solar power is directly dependent on the irradiance of a particular area. Thus, the selection of the location forms a main factor for the subsequent success in solar power generation. As we already know, irradiance is generally expressed in kWh/m^2 and indicates how much light power per square meter impinges on an area.

Global Radiation in Germany

Based on satellite derived values and ground measurements at DWD stations

Annual sum 2020

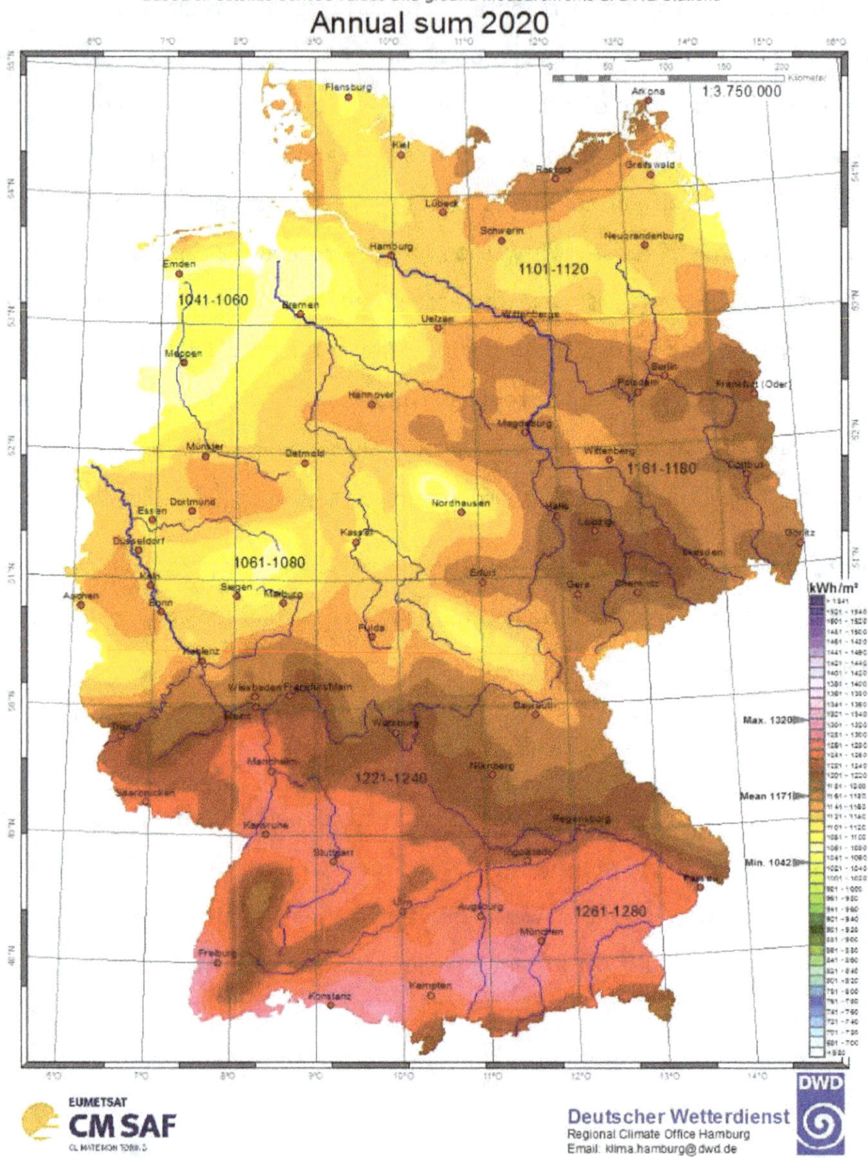

In order to determine the irradiance of a certain area, we can refer to the page

https://globalsolaratlas.info/ (or the tool "PVGIS" as in the previous chapter). We just have to enter the coordinates or the name of the area and the value of irradiance will be displayed.

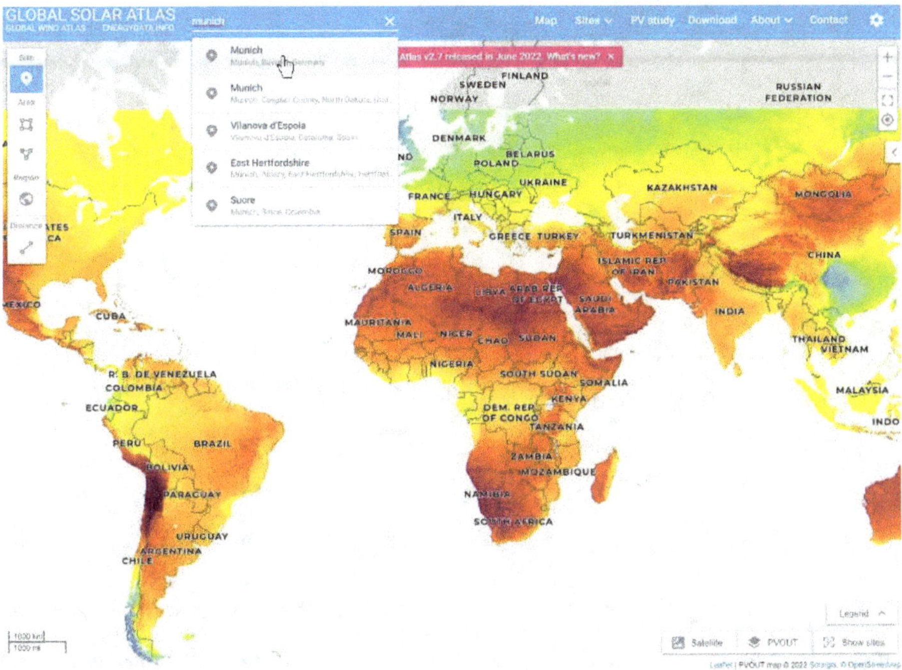

After entering the desired location and opening the sidebar on the right, we can view the desired information. The following parameters are displayed:

- Specific PV power

- Direct normal irradiation

- Global horizontal irradiation

- Diffuse horizontal irradiation

- Global inclined irradiation at optimum angle

- Optimal inclination of the PV modules, air temperature & altitude meters

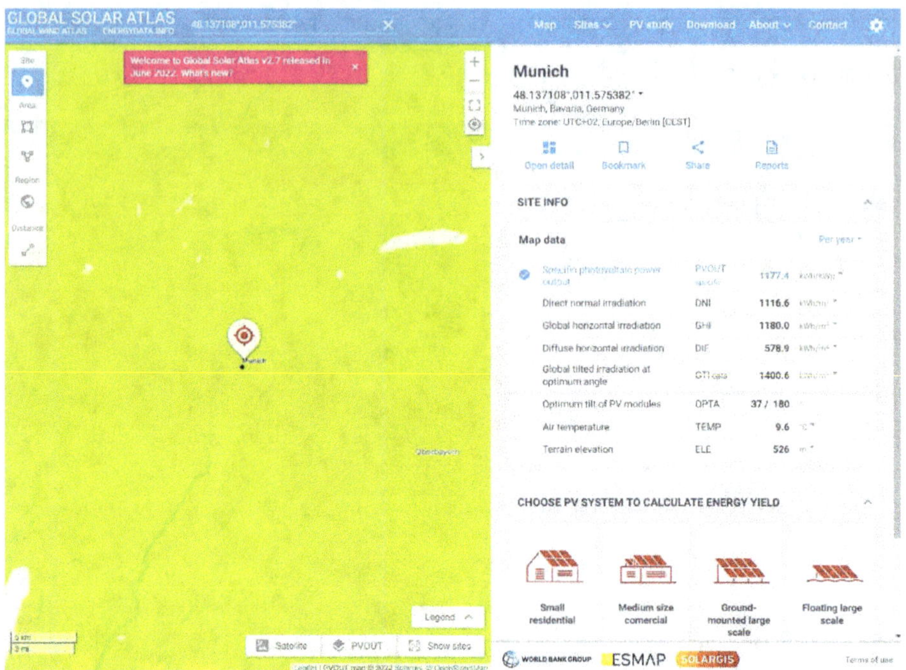

The irradiance varies according to the movement of the sun during the year. The angle of the sun varies between 23.5 degrees positive at the summer solstice and 23.5 degrees negative at the winter solstice.

Global radiation (direct + diffuse solar radiation):

Solar irradiance is divided into direct and diffuse solar irradiance and is referred to together with the term global irradiance. The irradiance changes throughout the year and depends on the weather and the location of the area. Diffuse radiation is caused by scattering of light by clouds or fog. Direct radiation, on the other hand, hits the earth's surface directly without scattering.

Inclination of the PV modules:

PV modules are tilted at a certain angle to take advantage of the maximum solar radiation. We have already dealt with this topic in the example of the stand-alone system in the previous chapter. For year-round operation of the PV system, we can

also check the ideal tilt of the PV panels using the website
https://globalsolaratlas.info/. The optimal tilt is shown to us as "Optimum tilt of
PV-modules". However, if the PV system is installed on the roof of the house, we
generally do not use the optimum angle and simply install the PV panels parallel to
the already specified roof pitch for various reasons (effort, cost, aesthetics ...).

We can also have this website calculate the average annual values that we can
obtain with a PV system at a given location. To do this, we select the use case in
the lower-right area, for example "Small residential" and in this way we obtain the
annual average values. By clicking on the small blue gear labeled "Change PV
system" we can adjust the initial values for the calculation.

PV SYSTEM DATA

PV system configuration

Pv system: **Small residential**
Azimuth of PV panels: **Default** (180º)
Tilt of PV panels: **Default** (37º)
Installed capacity: **1 kWp**

 Change PV system

Annual averages

Total photovoltaic power output and Global tilted irradiation

1.133

MWh per year ▾

1398.9

kWh/m² per year ▾

Open detail

6.3 Step 3: Check current consumption or load to be connected

After we have selected our location and checked it in terms of solar radiation, we now come to the third step of planning. This step is about calculating the power consumption of your household. This step is very similar to the previous example of the island system. However, there are a few differences.

We distinguish two cases for this. In the first case, we consider the planning of a photovoltaic system <u>without</u> battery storage and for self-use of the electricity. To do this, you need to list all the critical consumers (refrigerator, washing machine, electric stove, ...) that you want to operate with the photovoltaic system during the day. To do this, create a table with a column for the name of the device, with a column for the energy demand [W] and with a column for the duration of use [h]. After you have listed all the devices, note the number of hours per day that these devices are on during the hours of sunlight. Here, we only consider the time during the day because a photovoltaic system (without battery storage) only provides electricity during the hours of sunlight. The third step in the load calculation is to determine the wattage of each appliance based on the nameplate information and include it in the table. Now calculate the total energy demand in watt-hours by multiplying the wattage of the appliances by the hours required to operate them. This could then look like this, for example:

Device	Energy demand [W]	Service life [h]	Total energy demand per day [Wh]
Refrigerator	150	4	600
Washing machine	1500	3	4500
Electric stove	2500	1	2500
PC	100	3	300
...		Total:	= 7900 Wh = 7.9 kWh

After calculating the total load in watts and the total energy demand in watt-hours, we can estimate the capacity of our solar system.

In the second case, we consider the planning of a photovoltaic system <u>with</u> battery storage and for the self-use of electricity. In this case, we can make it easier for ourselves than in the first case. To estimate the total energy demand, we can simply take a look at our electricity meter. The electricity meter is located in the electrical cabinet of our house and counts our electricity consumption of the whole house in kWh. To do this, we note the current value on an average day, wait 24h and then read the value again. In this way, we quickly and easily get the total energy consumption for an average day in kWh. If we now multiply the difference of the two read values by 365, we get the average electricity demand for our household for one year in kWh. If you want to know more precisely, you can also take a look at an old electricity bill. Here, the value for one billing period (e.g., 1 year) is given. For the average electricity consumption per day, we can also divide this value by 365.

6.4 Step 4: Detailed planning

Now that we have dealt with the basic steps of the planning, we can start with the detailed planning of the PV system. For this, we use in the following the calculation program "PV*SOL" or "PV*SOL premium" of the company "Valentin Software". There is a 30-day trial version, which is long enough for non-commercial planning for your house. Download the software at the following address: https://valentin-software.com/en/downloads/. There are also many tutorials about this software.

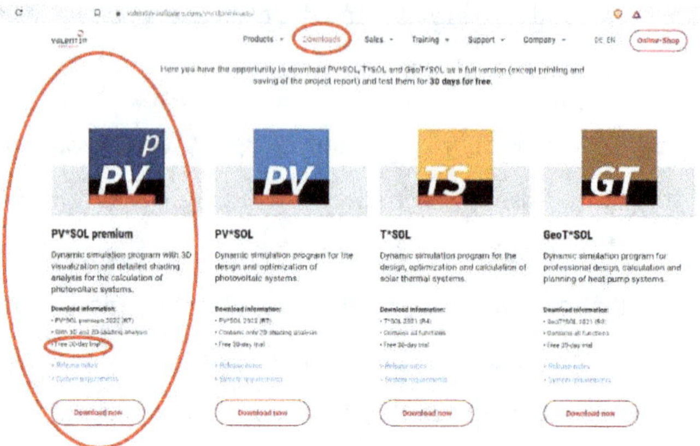

This software will now guide us step by step through the detailed planning. After we have started the software, we will see several buttons in the menu bar in the upper area, each representing a step of the planning. We start on the far left and work our way to the right. The first button, however, is only for the start screen, in which you can see news about the software, sample projects, call up saved projects and create new projects.

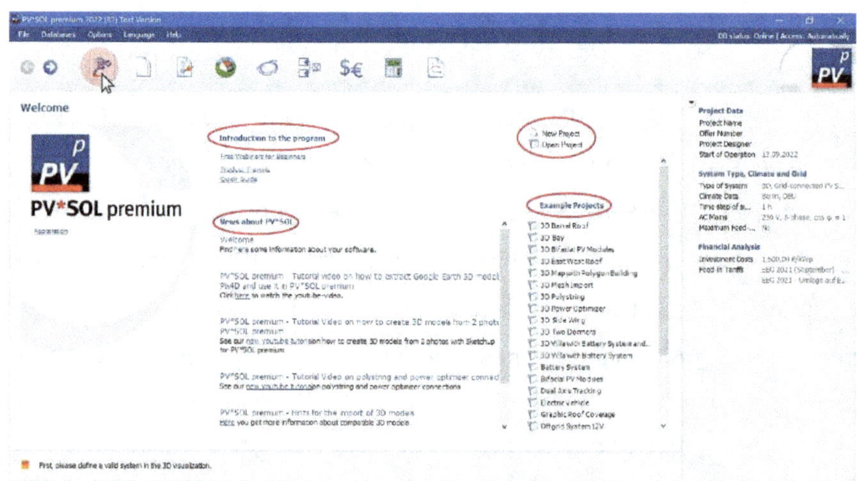

Therefore, we start with the second button. Here, we can enter a project name and a project description. However, we do not need customer details or quote number in this case.

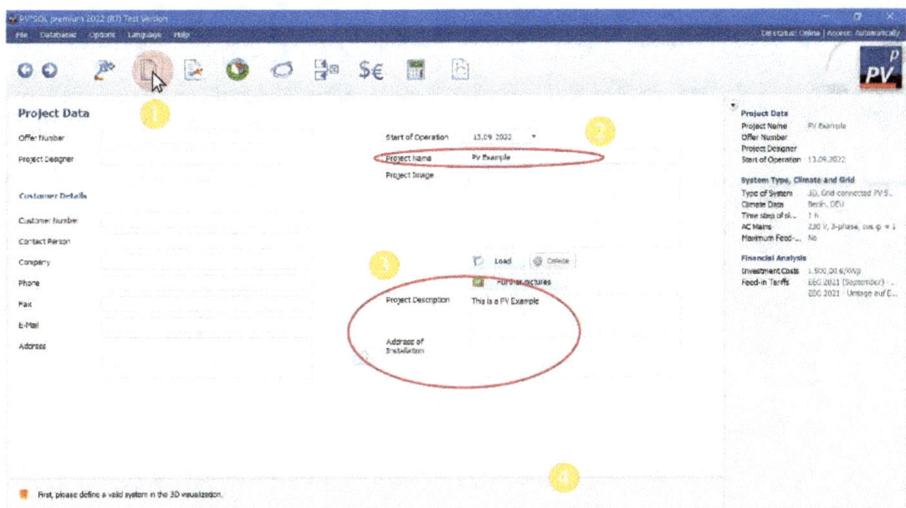

Then we switch to the third button. Here we can set the system type, specify the location and the power grid.

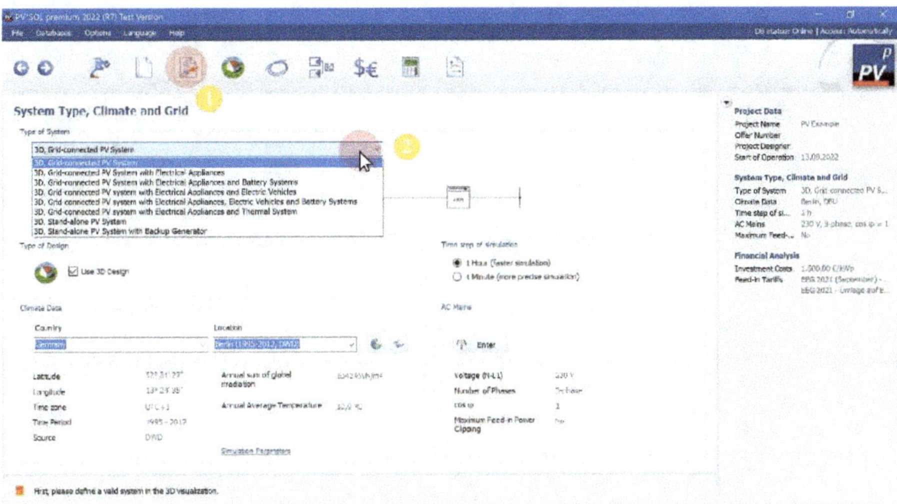

For example, we select a grid-connected PV system with connected electrical appliances (consumers), i.e., the type "3D, Grid-connected PV-System with Electrical Appliances" in the drop-down menu. If we also want battery storage, we would select the same option with the addition of "... and Battery Systems".

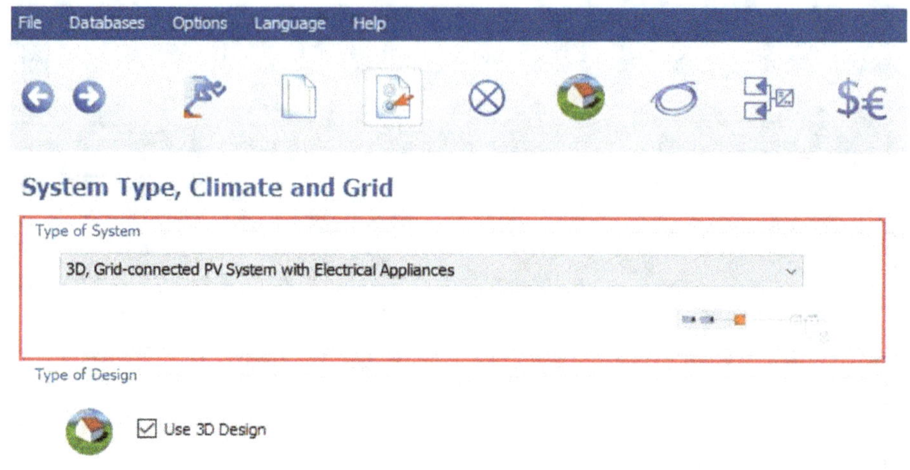

System Type, Climate and Grid

In the lower area we add the country and the location **(1) of** the PV system, and on the right side **(2) we** check if the information about the power grid is correct. If the data are not correct, we can edit them by clicking on "Enter".

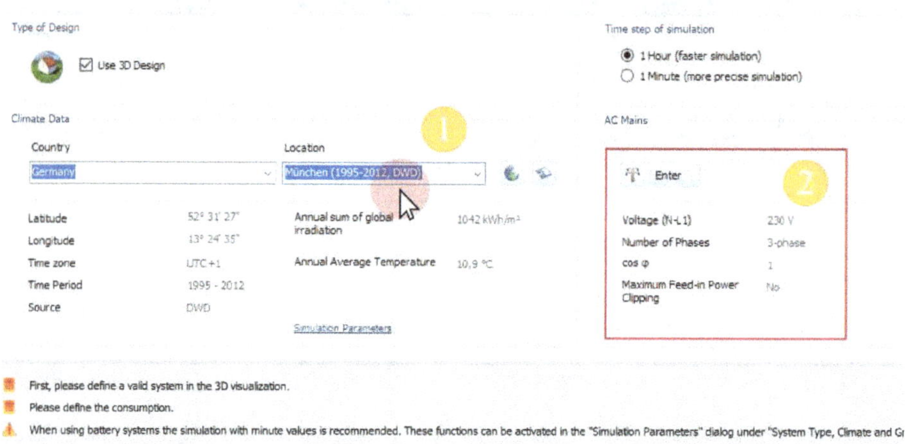

In the next step **(1)** we have to enter our electricity consumption. We do this by clicking on the drop-down menu "Add consumption" **(2)** and selecting the option "Load profiles / individual appliances". By the way, in the right area (framed in red) the previous entries or the placeholder values for our project are displayed for all steps.

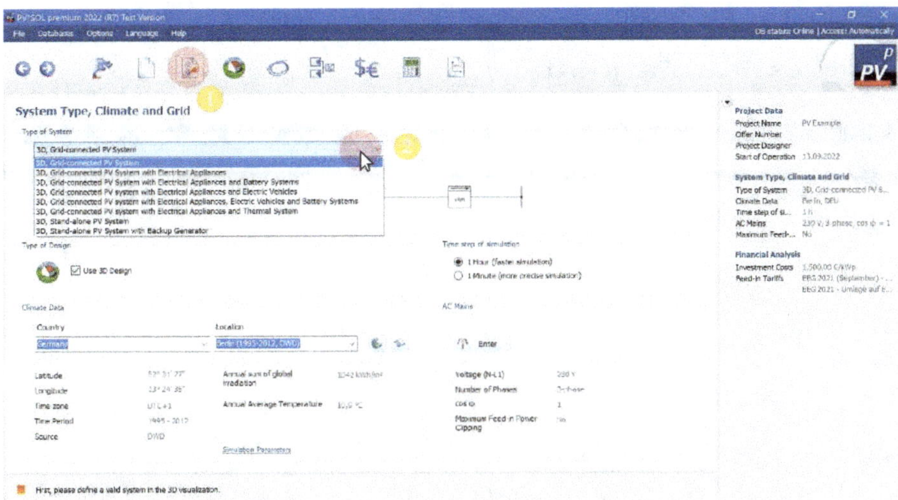

A window will open in which we select the scenario that best describes our household, e.g., a household with 2 adults and 2 children **(2), in the** option "Load profiles (from measured values)" (**1**).

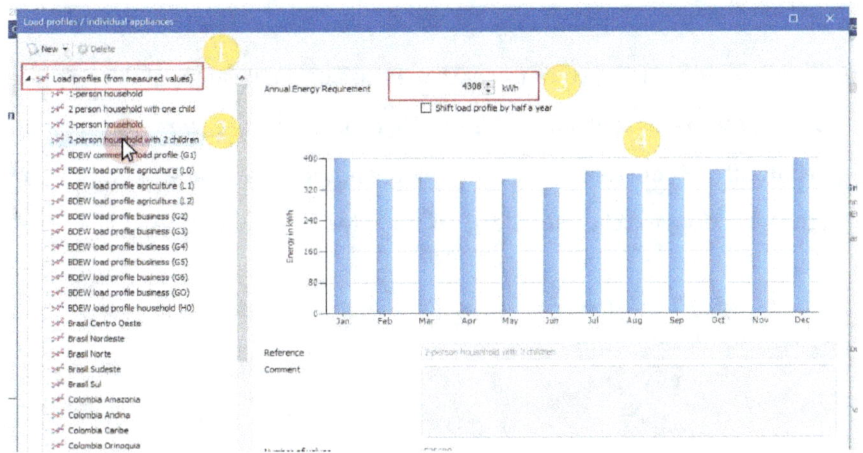

In the field "Annual Energy Requirement" a default value is entered, which we can replace with the value we have read or calculated for our home based on the electricity bill. The program splits this annual consumption – depending on the load profile – as shown in the bar chart **(4)**. In our case, for example, we leave the 4308 kWh.

In the next step, we capture the design of our house and the PV system. To do this, we create a 3D visualization by clicking on the button "Edit".

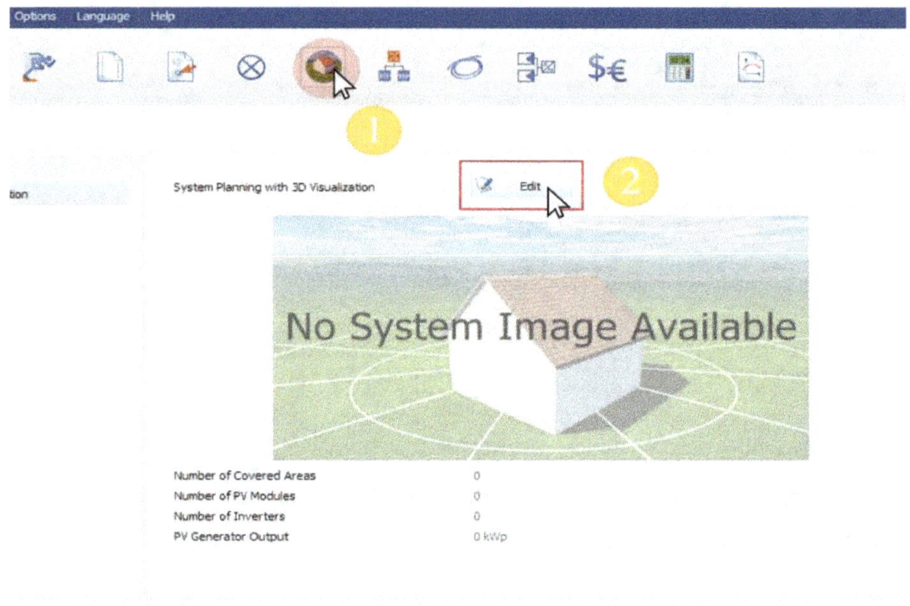

A window opens in which we can select the house or the location of the PV system. In our case, for example, it should be a gable roof. Therefore, we select the option "Gabled Roof". Of course, you can also select another roof shape or an open area. After the selection, we click on "Start" to start the 3D visualization.

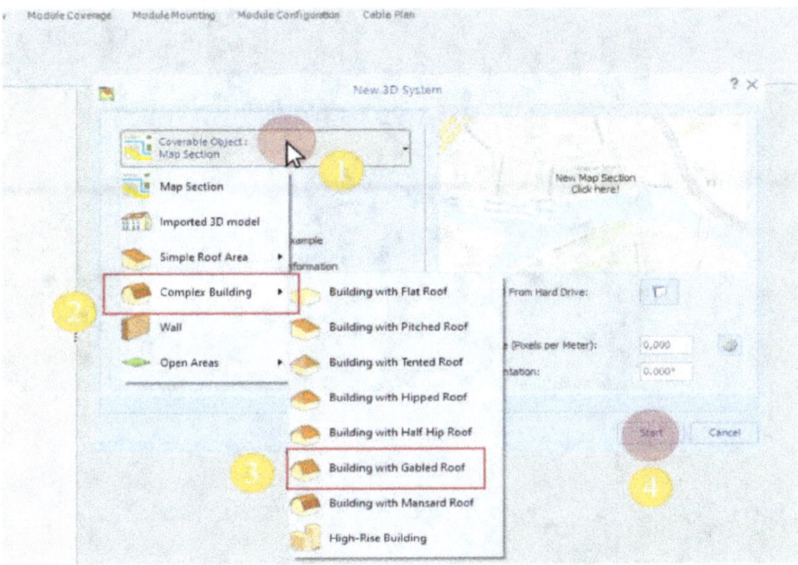

In the 3D visualization, we get the model of our house. We can even add more houses, but also trees, or walls with the buttons from the upper area to make the planning as realistic as possible. Simply drag and drop the desired elements into the display area. In this case, however, we assume a freestanding house.

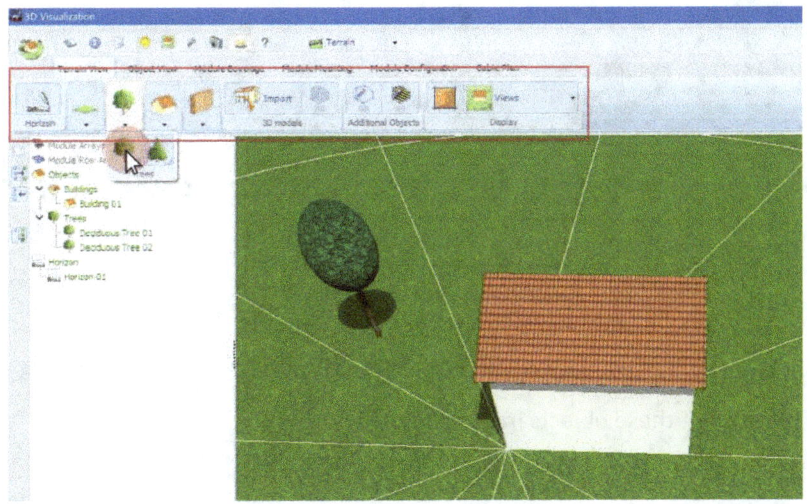

By right-clicking on the house and selecting the option "Edit" we can adjust the footprint of the house or the roof area and the roof pitch.

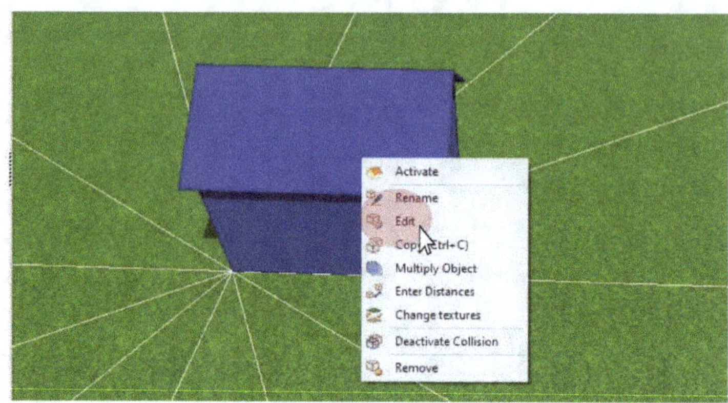

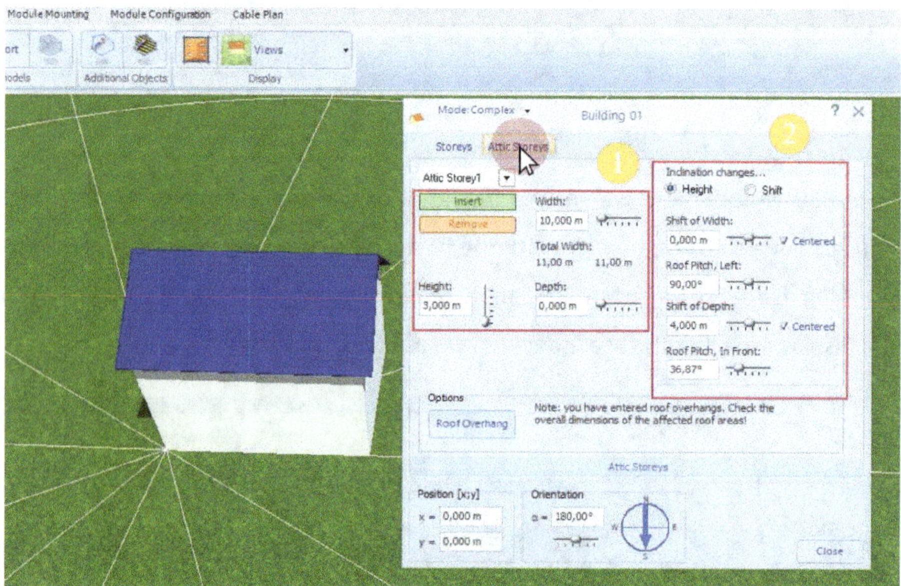

By clicking the button "Close" we finish the settings again.

By selecting the button "Object View" in the upper area, you can zoom directly to a roof surface and add roof elements, such as chimneys, skylights, etc... We will also do without these objects in this case.

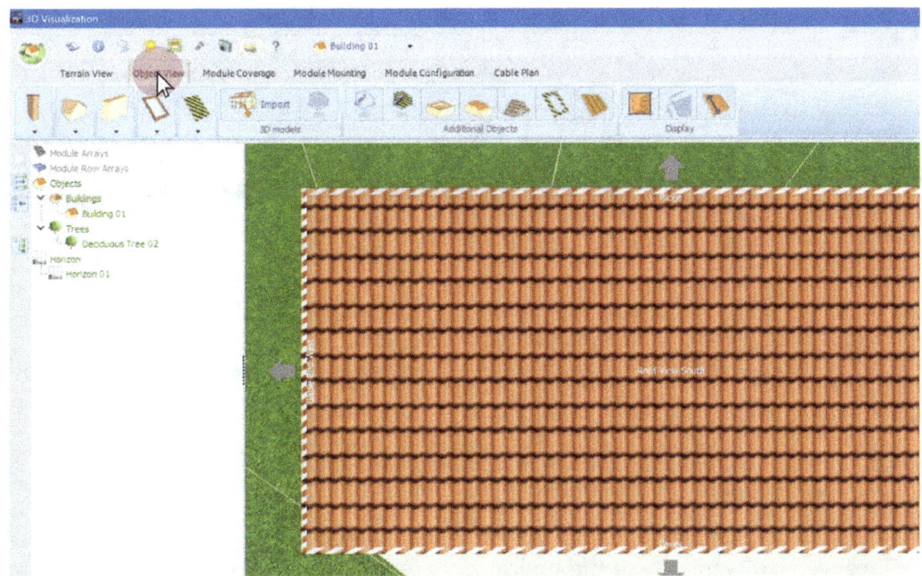

In the next step, we can add PV modules to our virtual roof. We do this by selecting the menu "Module Coverage" and clicking the button "New Module".

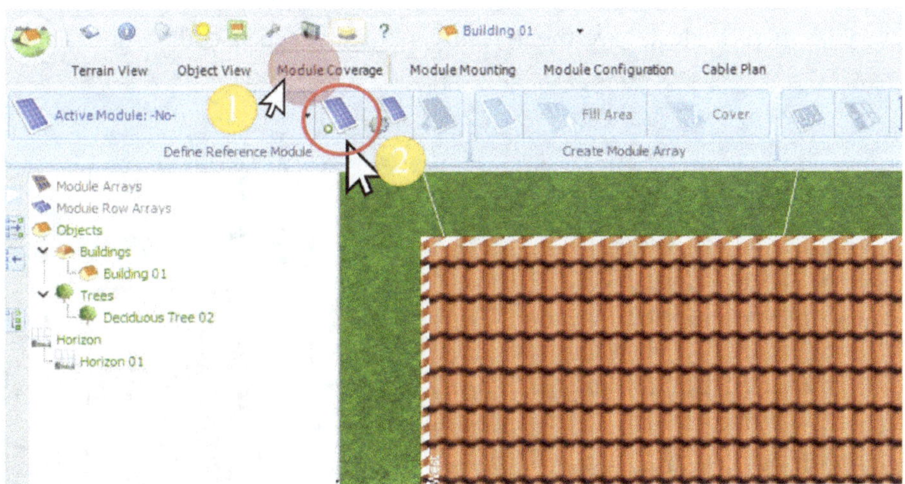

A window opens in which we can select the desired PV module of various manufacturers (framed in red) from a database. In this case, however, we simply select a standard monocrystalline PV module with 200 Wp.

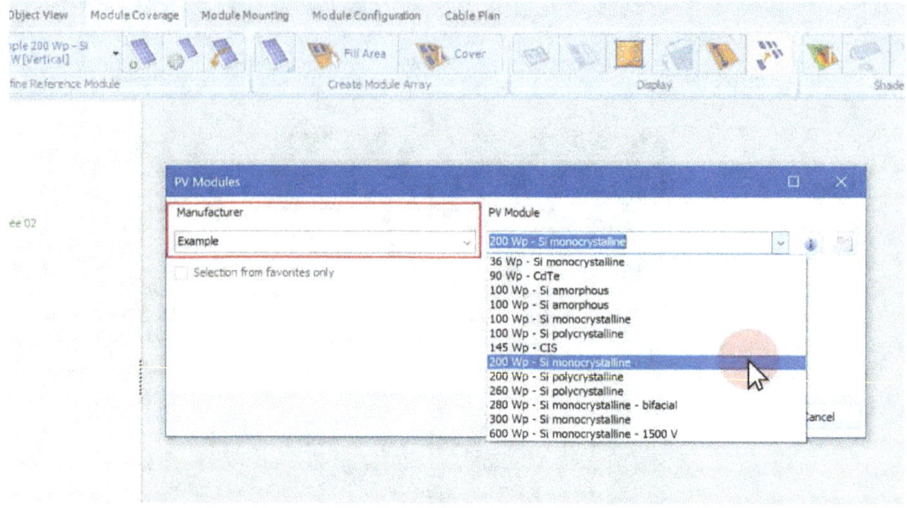

Since we now want to use the entire roof area to generate as much electricity as possible (self-consumption + surplus feed-in), we can select the button "Fill Area" and use the PC mouse to draw an area in which we would like to place the PV modules (e.g. the entire roof). But first we have to set the distances between the modules (here we just leave the default parameters) and select the "Installation Type", e.g., the option "Flush Mount – good rear ventilation".

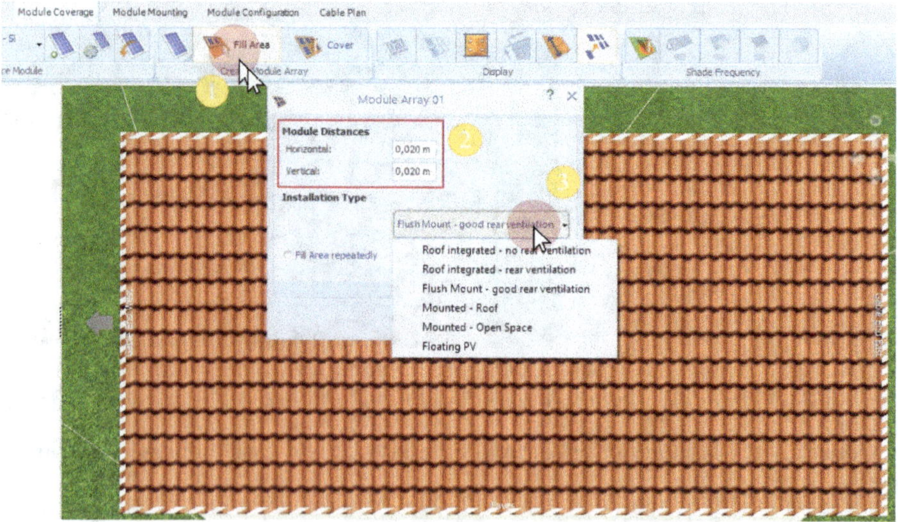

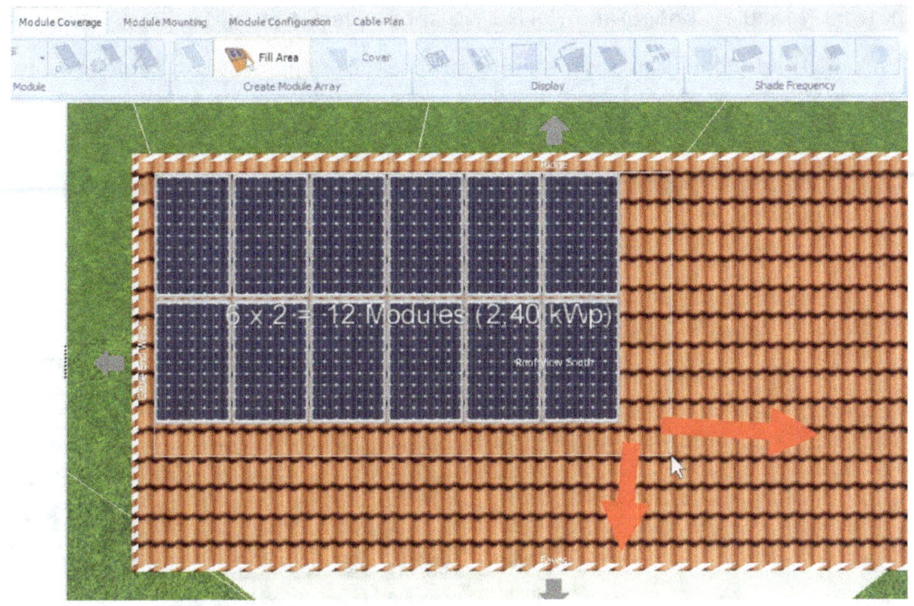

We then confirm with "Ok" so that the PV array is placed:

Then we could plan the mounting of the modules by clicking on "Module Mounting". However, we will skip this step in this example. The next step is to configure the wiring and select the PV inverter. We can do this by clicking on the button "Module Configuration" **(1).** We then select the button "Configure all Unconfigured Modules in this mounting surface" **(2) in** this area to be able to select an inverter for the PV modules.

To do this, we select the button "Inverter-Selection" in the window that opens.

Another window opens. Here we first select a preferred manufacturer, e.g., the company "SMA Solar Technology AG" **(1)** and then select all inverter types of the company from the database for checking **(2).**

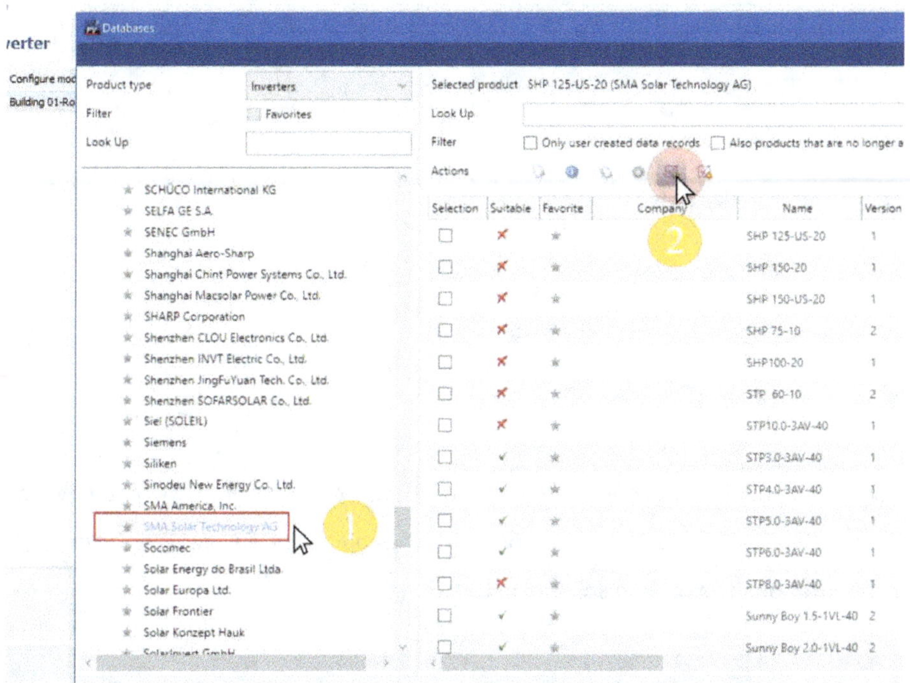

Then we close this window by clicking the button "Select".

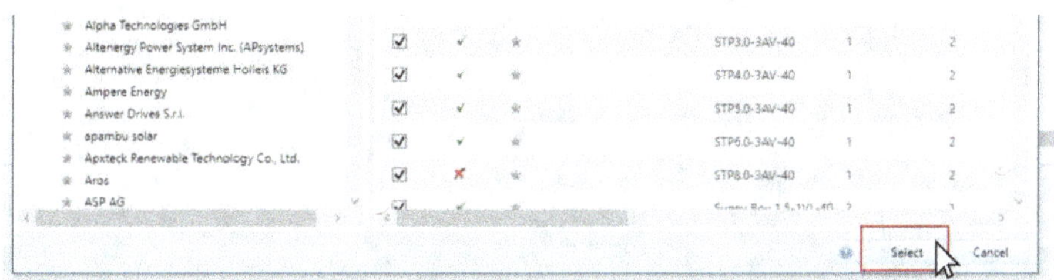

Next, we click on "Suggest Configuration" **(1)** so that the program automatically selects the ideal circuit and the ideal inverter of the company "SMA AG" from the multitude of possibilities. We see the result after a short calculation, then in the lower area **(2)**. With a click on "Select Configuration" we could also select the configuration manually from numerous calculated possibilities.

If we want to select a configuration manually, we would have to click on the button "Start" **(1)** after "Select Configuration" in the opening window and then all possible configurations would be displayed in the area **(2).**

However, we close this window with the button "Cancel" and remain with our automatically generated configuration. Here we can check the automatic selection with a click on one of the checkmarks or on the button "Check System" in the lower area.

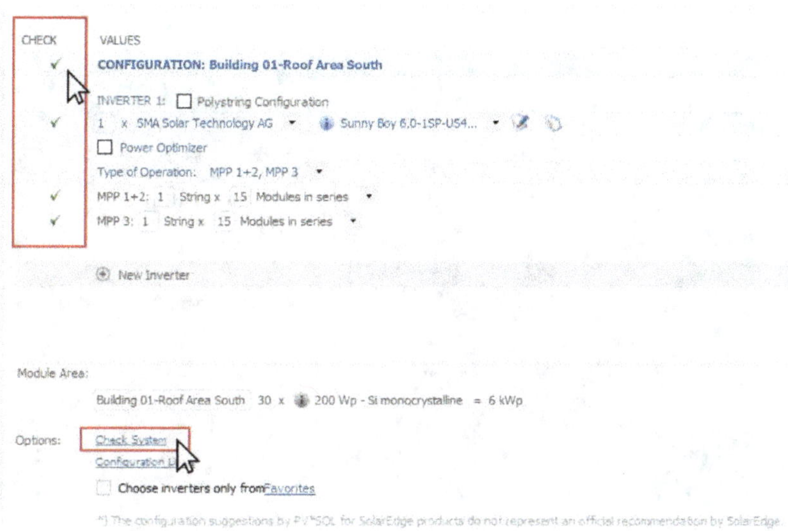

We can see for ourselves in the following window that the inverter is designed large enough and that all values remain in the green range.

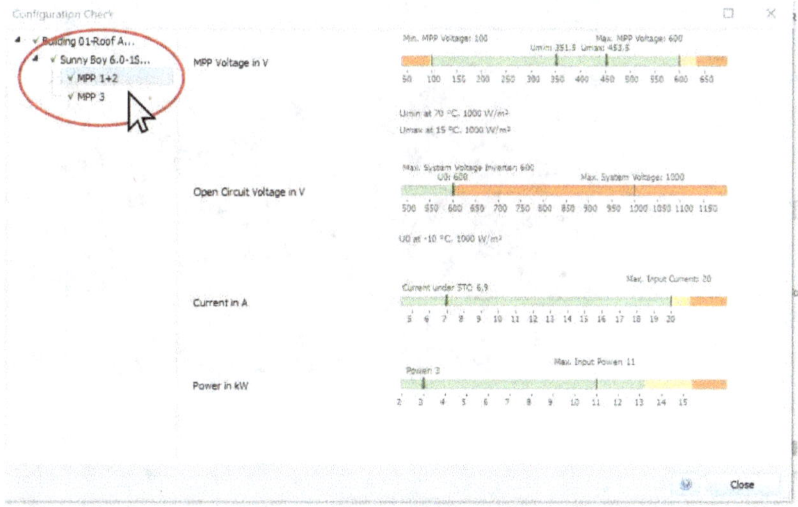

Use the "Close" button to return to the initial window.

Here we confirm with the button "Ok" in the lower area. Now the division of the modules into two serially connected "Strings" per MPP tracker of the inverter is displayed (red and orange):

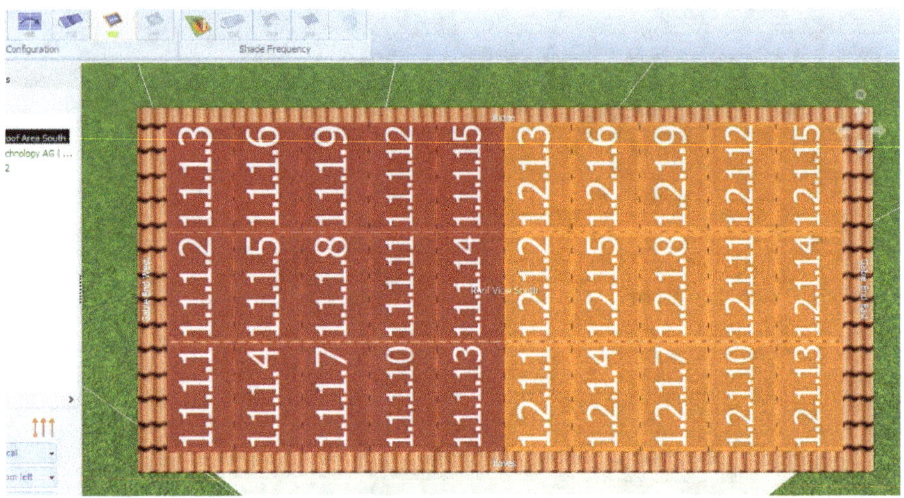

With the last menu item "Cable Plan" we can display the cabling of the PV modules or plan the further cabling up to the house.

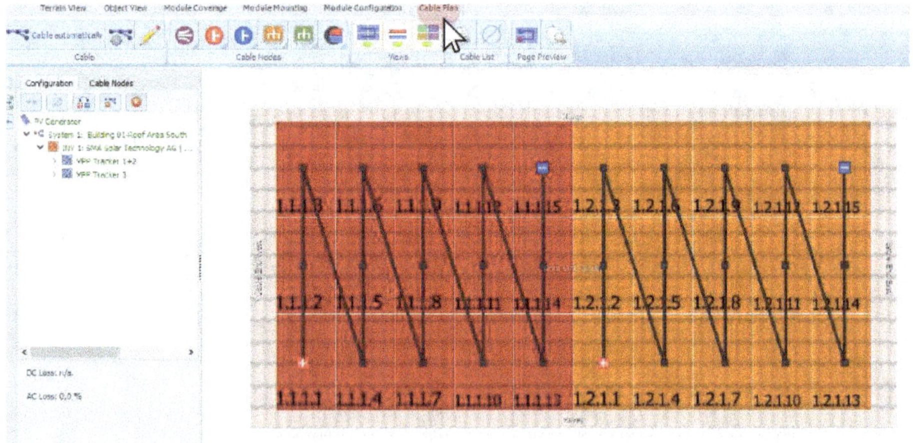

Then we are done with the 3D visualization, we can switch back to the view "Terrain View" and then close the window. It is important that we now select in the following pop-up window that the data is transferred to "PV*SOL" so that the previous work is not lost.

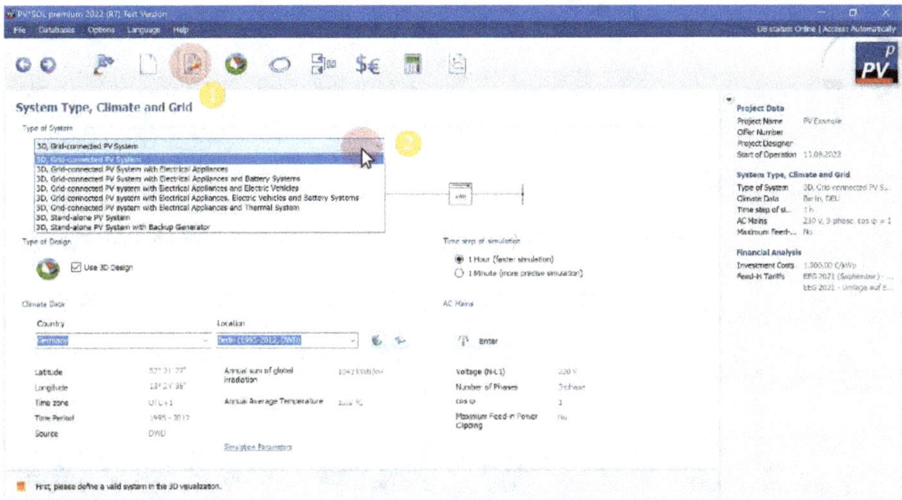

We are then asked whether we want to perform a shadow analysis. This is generally useful, but in our case we don't have any element that could create a shadow, so we can select the option "Ignore Shading".

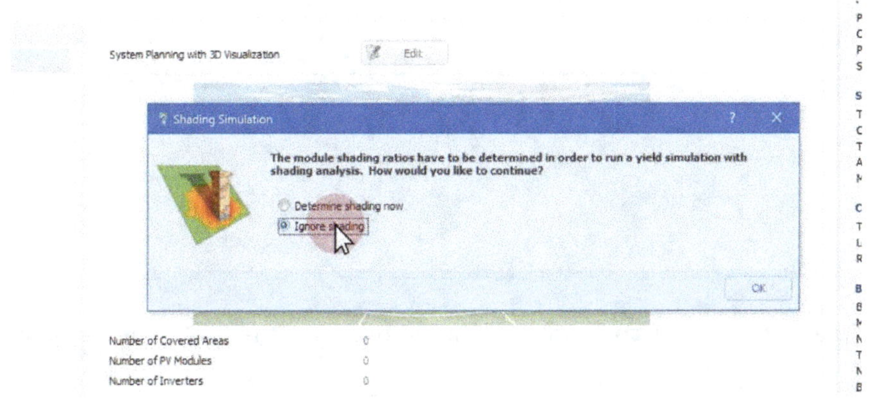

In the section "Cables" we can then create the final wiring diagram of our PV system, i.e., add a solar meter, fuses, etc.

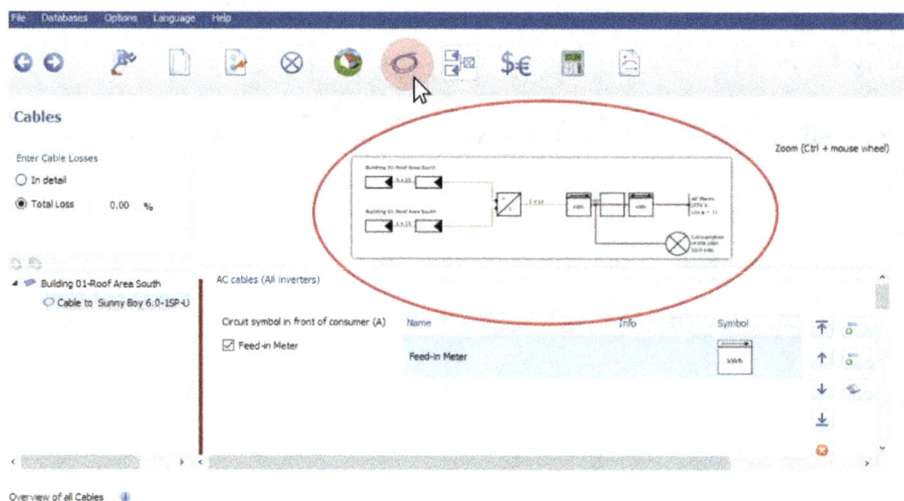

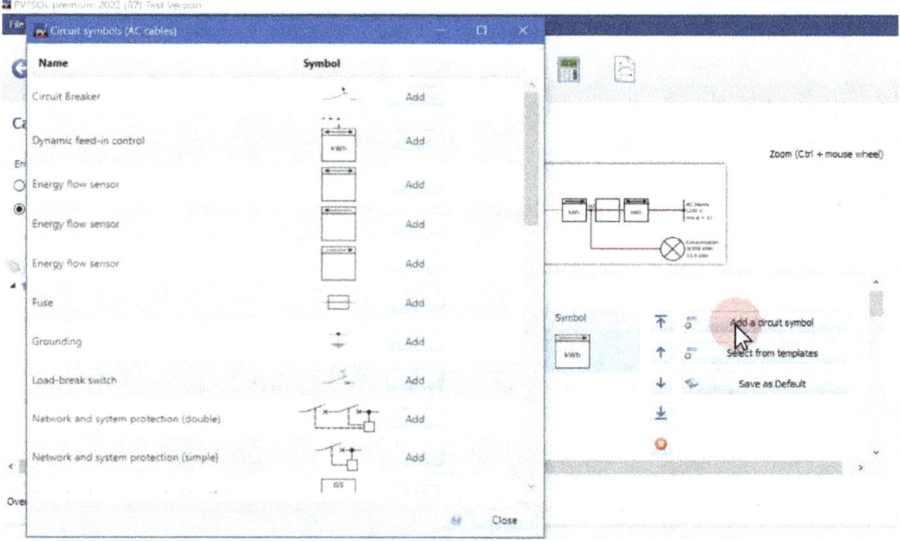

In the section "Plans and parts list" on the one hand the final circuit diagram is displayed and on the other hand the required components are listed in a parts list by clicking on "Parts list".

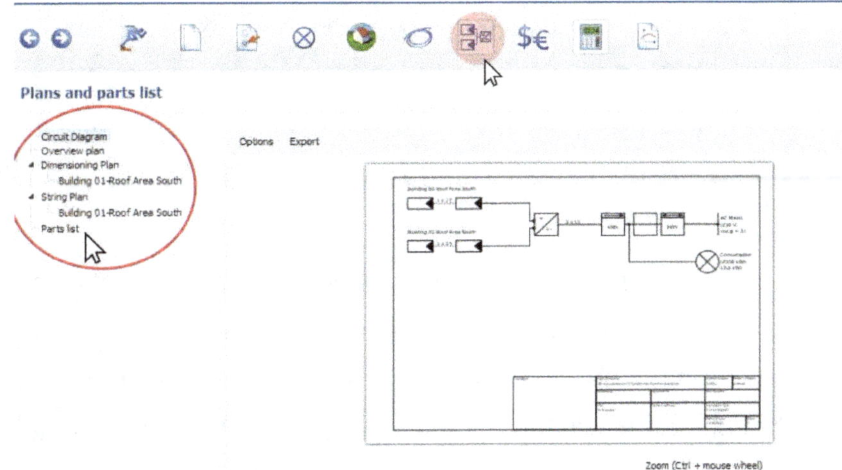

Zoom (Ctrl + mouse wheel)

We will skip the economic analysis (button "Financial Analysis") in this example. Here you are welcome to try it yourself. As a last step, we rather want to display the results with the button "Results". After a calculation we see the produced PV energy, the own use and the grid feed-in depending on the graph selection clearly displayed in the course of the months.

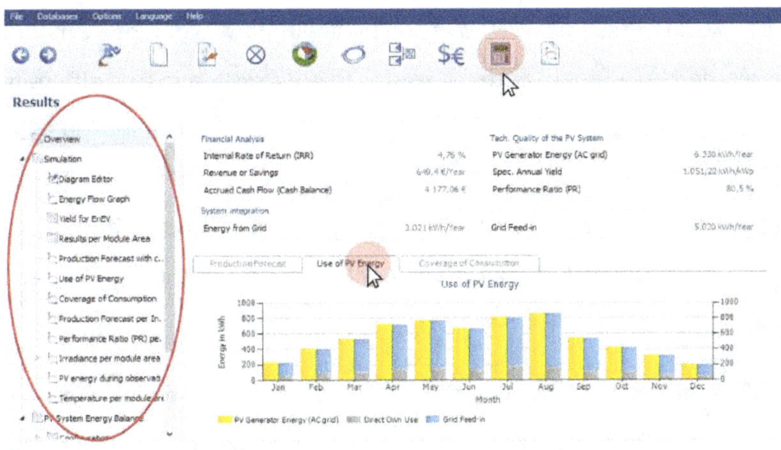

Closing words

Excellent! You have made it. Congratulations if you have made it this far. This chapter ends the photovoltaic book.

Together in this book we have explored the basics of photovoltaic technology as well as learned how to plan and install an on-grid as well as an off-grid photovoltaic system with or without battery storage. Now it's time to plan your own PV system for your house, garden shed, Tiny House, RV, garage, workshop, car, camping, ...,.

If you are still unsure about a few points, you should definitely get professional help from an electrician for your individual project.

However, by now you should have mastered the basics in any case. For example, in this book we learned what the abbreviation Wp means, whether you should choose monocrystalline or poly crystalline PV modules, what you need an inverter for and how to choose one. We also thought about how to mount PV modules, as well as learned how to add battery storage to a PV system in two different ways (DC-coupled or AC-coupled). We also looked at the parallel connection and series connection of PV modules and battery storage and their effect on current and voltage and many other details step by step. So, we've covered quite a bit! Be sure to take a look at the last pages for books on similar topics (e.g., electrical engineering)!

If you liked this book, I would personally be very happy if you leave me a rating and a short feedback, as well as recommend the book! Especially, a rating will also help other interested parties in their decision. Thank you very much!

Books on topics you might also like

All books are available online on the usual sales platforms. It's best to just search for the title, or feel free to visit my author page. Some of the books may not be published yet and will be released or found soon. Take a look at the books of your choice and your copy as e-book or paperback!

3D Printing:

CAD, FEM, CAM (3D Object Creation, Design, Simulation):

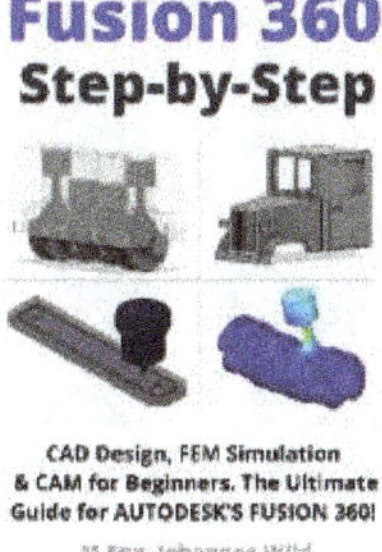

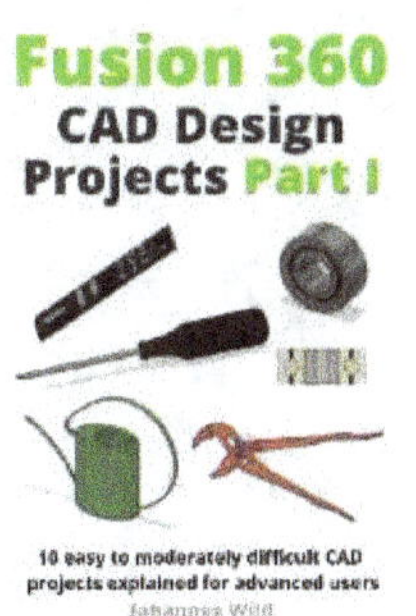

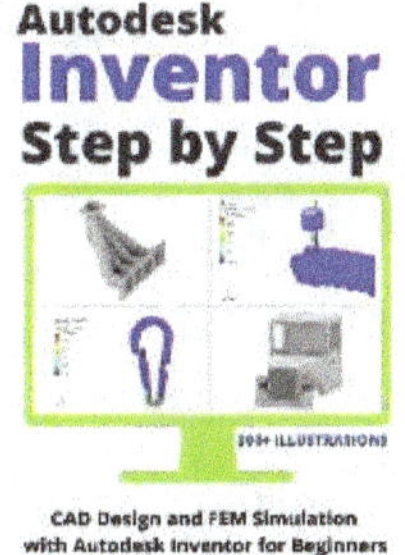

Electrical Engineering:

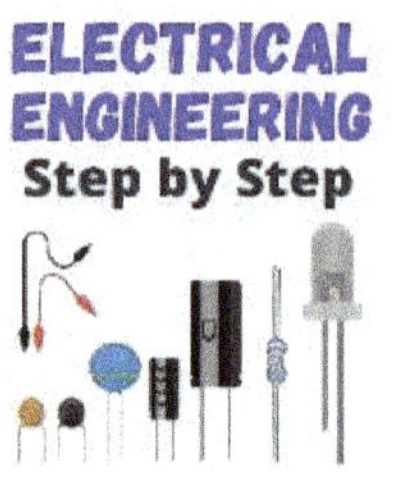
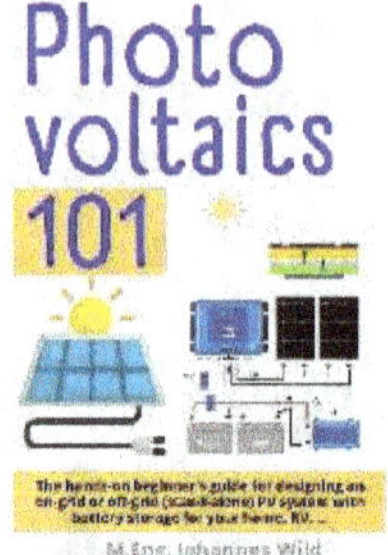
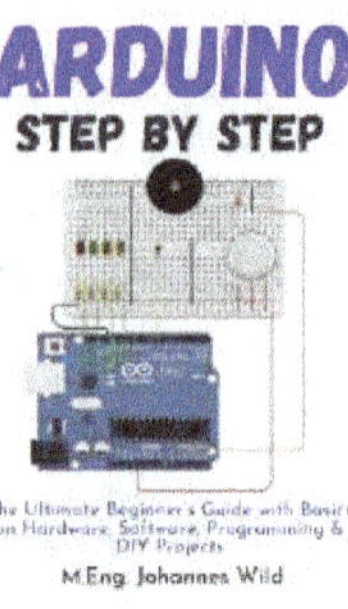

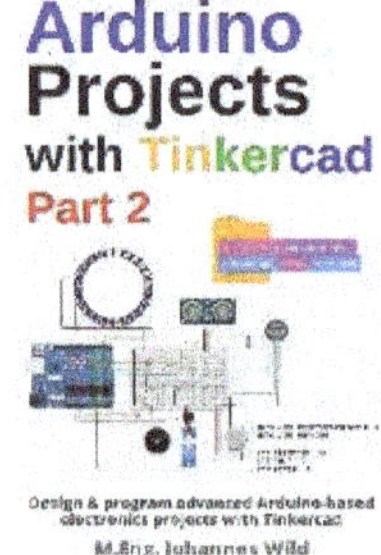

Programming and other Software:

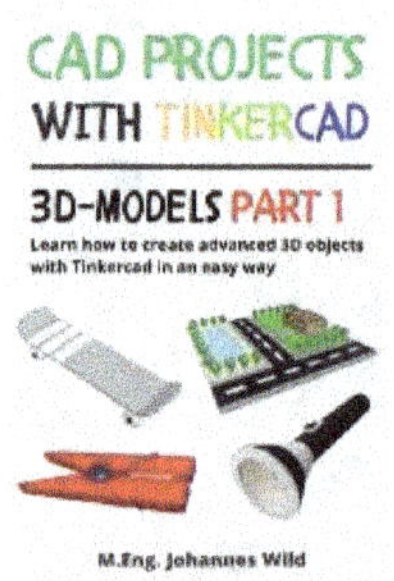

There are also identical video courses for some of these books:

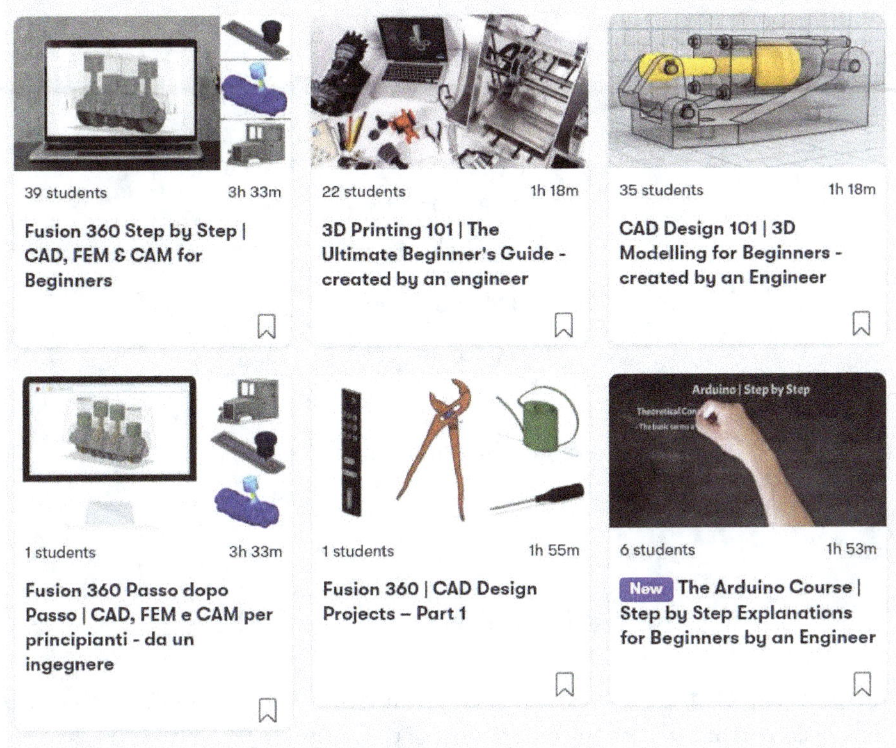

They are hosted on the learning website: skillshare.com

Be sure to use my following friends & family referral link to get a month of membership for free !

(I will get a little bonus if you choose to stay, so we will be both happy. Thanks in advance!)

https://www.skillshare.com/r/profile/Johannes-Wild/854541251

It is best to copy the link in your browser to access the free month !

Sign up today and deepen your knowledge!

Imprint of the author / publisher

© 2023

Johannes Wild

c/o RA Matutis

Berliner Straße 57

14467 Potsdam

Germany

Email: 3dtech@gmx.de

This work is protected by copyright

Thank you so much for choosing this book!